Marine Conservation Society

D1434982

MARINE FIELD COURSE GUIDE 1

ROCKY SHORES

PORT ERIN MARINE LABORATORY
1892 – 1992

STEPHEN J. HAWKINS
Port Erin Marine Laboratory
University of Liverpool

HUGH D. JONES
Department of Environmental Biology
University of Manchester

Typesetting by Shirley Kilpatrick, Icon Publications Ltd, Kelso, Scotland

Illustration by Ursula Klinger, Ireland

Design by Jane Stark, Connemara Graphics, Ireland

British Library Cataloguing in Publication Data
Hawkins, S. J.
    Marine field course guide.
    1, Rocky shores. S. J. Hawkins and H. D. Jones
    1. Marine Organisms. Ecology
    I. Title II. Jones, H. D.
    574.52636

    ISBN 0-907151-58-2

IMMEL Publishing
20 Berkeley Street
Berkeley Square
London W1X 5AE

Telephone: 071 491 1799 & 071 493 9911
Fax: 071 493 5524

CONTENTS

Acknowledgements   4

Preface, Dr Bob Earll, Marine Conservation Society   5

Introduction   7

The Shore Environment   9

Patterns of Distribution on Rocky Shores   13

Life Histories of Shore Organisms   23

Identification Section: 25
    Putting names to intertidal organisms   27
    Lichens   29
    Green Algae   31
    Brown Algae – Wracks   33
    Brown Algae – Kelps   35
    Other Brown Algae   37
    Red Algae – More exposed shores   39
    Red Algae – More sheltered shores   41
    Sponges   42
    Hydroids   43
    Anemones   45
    Tube Worms – Serpulids   47
    Barnacles   49
    Bryozoans   51
    Other Prominent Space-occupying Species   53
    Limpets   55
    Topshells and Dogwhelks   57
    Winkles   59
    Other Important Grazers and Predators   62

Colour Plates   65

Surveying Rocky Shores: General principles   81

Class Exercises   93
    I   Indicator species and species diversity   94
    II   Semiquantitative exercise on zonation   96
    III   Exercise on minimal sampling size   98
    IV   Quantitative zonation exercise   101
    V   Limpet population and shape   102
    VI   Dogwhelk shell shape   104
    VII   Fucoid morphology and wave action   106

Energy Flow in Shore Communities   108

Causes of Distribution Patterns   110

Dynamics of Shore Communities   116

Project Suggestions   122

Long-term Studies and Experiments   125

Conservation   130

Bibliography   132

Glossary   142

ACKNOWLEDGEMENTS

This book grew out of class sheets and lectures prepared for University of Manchester students attending marine ecology field courses. We are very grateful to the many undergraduates whose patience, curiosity and acute questioning sharpened our wits and helped improve the material. We also thank the many postgraduate students and colleagues who have helped us teach these courses and similarly helped improve the final result. We have also used this material for extra-mural evening and residential classes, and again we thank the many students for their comments.

We are grateful to many colleagues in both the University of Manchester and the University of Liverpool for their advice. Dr Richard Hartnoll, Dr S. E. R. (Bill) Bailey and Dr Peter Gabbutt deserve special thanks for the development of some of the projects we have outlined. We thank the following for advice on parts of the book: Dr Gray Williams, Dr David Reid, Dr Richard Hartnoll, Dr John Bishop, Dr John Thorpe, Dr Liz Sheffield, Dr George Russell, Dr Martin Wilkinson, Professor Trevor Norton, Dr Mark Ward. Thanks are also due to Professor A. J. Southward for his general help and guidance to S. J. Hawkins over the years. We alone are responsible for any errors or inconsistencies remaining and would be grateful if they were pointed out to us. We also thank Dr Martin Attrill, Dr Elaine Fisher, Dr Nicola Manley, Dr Andy Hill and Mrs B. Brereton for their help in preparing the manuscript for us. S. J. H. would also like to thank Dr Elspeth Jack for her great patience and help on the shore over the last ten years.

Dr Bob Earll of the Marine Conservation Society, who motivated us to write this book, deserves special thanks. We thank also the publishers, Immel Publishing, particularly Dr Peter Vine and Jane Stark for their patience and advice. Thanks also to Ursula Klinger for the line drawings. Most of the photographs were taken by one or other of the authors. Other photographs are credited in the Plate legends.

Much of the field work on which this book is based was supported by various grants to S. J. Hawkins from the Natural Environment Research Council and the Nuffield Foundation.

The public's awareness of the environment and the issues that threaten it continues to grow. Telling people about the natural splendour and complexity of the marine environment remains a major challenge and yet the seashore presents one of nature's most inviting opportunities to study marine life. For many years we have all been confronted with very traditional approaches to field guides with their keys which proved very difficult to follow and meant that specimens continuously had to be collected and brought in to the laboratory. IMMEL has pioneered innovative approaches to species identification in the marine environment. This valuable guide carries on that innovative tradition. It will enable students and teachers to identify the dominant British intertidal marine life with ease and on the shore, without the need for damaging collections.

The underlying idea for the book has been developing for many years. When I studied at Manchester in the 1970s, Hugh Jones was using tables of features for common intertidal groups to enable course students to compare and identify species with great ease. The core of the idea for the book was borne. This has been combined with excellent photographs and illustrations to enable students to tell species apart on the shore. When Steve Hawkins came to Manchester the collaborative venture blossomed to include the wider elements of seashore ecology.

An understanding of the environment is more than just knowing the names of species. The seashore provides excellent opportunities for students to come to terms with many ecological principles which will help them to appreciate the complexities of the environment. In the second part of the book a wide range of ideas for studies has been put forward which should greatly assist teachers of all groups to formulate practicals which will help develop ecological understanding.

It gives me a terrific thrill to see this book come to fruition and I'm sure it will excite the interest of new generations of people in the sea, its ecology and conservation.

by Dr Bob Earll
Head of Conservation
Marine Conservation Society

The seashore gives most people their first taste of the diversity of life in the sea. It is very much a boundary zone. The transition from a fully aquatic marine habitat to fully terrestrial conditions occurs within a few hundred metres at the most and often less than 10 metres. Therefore any shore is a sharp **environmental gradient** which is ideal for ecological studies – far more convenient than, for example, ascending a mountain which would provide a similar range of environmental conditions. Rocky shores are particularly suitable for research and education. They are essentially two-dimensional and are easily sampled **non-destructively**, without the need to dig and sieve which is so necessary on sandy and muddy beaches. Also most of the plants and **sessile** animals compete for a clearly defined **resource** – space.

This book has been written very much with ecological education in mind. It is based largely on material prepared for a variety of field courses at both introductory and advanced undergraduate level and for extra-mural adult education classes. Our viewpoint is that ecological research and practical exercises are worthless without accurate identification of the organisms involved. Therefore, our prime aim is to allow rapid and accurate non-destructive field identification of the common and ecologically important plants and animals found on a rocky shore. The species included are those organisms most likely to be encountered in transects of open rock surfaces and in shallow rock pools in the British Isles. Species which are abundant in crevices, under stones and amongst or on seaweeds are deliberately limited to a selection of the more obvious ones. There are many others but their study

invariably requires their removal and examination in the laboratory. The reader can identify these species using the more comprehensive pocket guides such as the Collins Guide (Barrett and Yonge [B & Y], 1972) and the Hamlyn/Country Life Guide (Campbell [C], 1976). A recent book by Hayward (1988) is excellent for many of the animals associated with seaweed. Where appropriate, the reader is referred to these books in the following way (e.g. B & Y, p.12; C, p. 34). Our book is not meant to replace these guides, but to complement them.

An illustration in colour of each species selected is shown in a block of plates, plus drawings showing the identification features. We have not used **dichotomous keys**; tables of identification features opposite the drawings have been given instead. We have tried to be as **taxonomically** accurate and up to date as possible. Where difficulties, uncertainties or controversy exist, the reader is given more advanced references and a recommendation about an acceptable level of accuracy. It is hoped that the important species can be identified easily and correctly.

The second aim is to provide interesting ecological information about rocky shores and the individual species. Therefore, the book starts with a short description of the intertidal environment and distribution patterns and outlines life histories. This leads to the identification section. Ecological notes and references to work on individual species are given throughout the identification section.

The third aim is to encourage stimulating conservation-oriented practical exercises and projects. This section is intended primarily for teachers of

field courses. Once an appreciation of the diversity of shore life and methods of study has been gained, the final aim is to outline the functioning of shore **communities.** Therefore energy flow and causes of distribution patterns are considered. Some illustrated examples of long-term changes and **biological interactions** on the shore are given to show the limitations of one-off descriptive surveys and to give an idea of the dynamic nature of shore communities. This is followed by some ideas for projects, long-term studies, and a couple of examples of ecological experiments which will cause minimal damage to shores. This part of the book is slanted to research and has more references to allow the advanced reader to go further. Guidelines on conservation of shores and a bibliography finish the book. Where a potentially new word or phrase is encountered for the first time it has been printed in bold and defined in a glossary of terms on p. 142-143.

The "British Isles" has been widely interpreted to include Ireland, the Channel Islands, the Isle of Man, Orkney and Shetland. As a result, some species such as the sea urchin *Paracentrotus* are only found rarely in the UK, but are exceedingly common on the Irish coast. Similarly, the topshell *Gibbula pennanti* is very common in the Channel Islands. In consequence, the book can also be used on the whole of the French side of the Channel, eastwards to Denmark, the west coast of Sweden and southern Norway.

We hope the book can be used by older school pupils and teachers, interested amateurs, conservation groups, and undergraduates. It should also be of some use to postgraduates and more advanced workers.

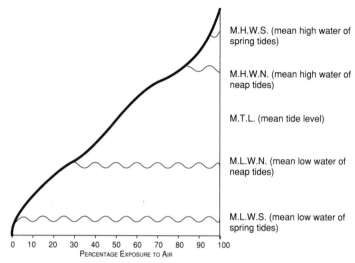

Fig. 1a. Gradient of marine to terrestrial conditions from low to high on a shore, based on % emersion time (time out of water) calculated from tide tables. (*After Lewis, 1964*).

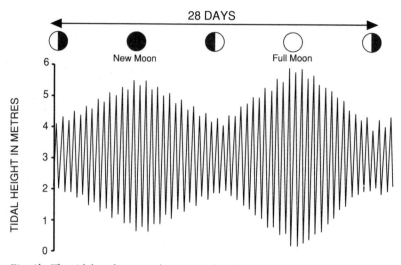

Fig. 1b. The tidal cycle over a lunar month. Note that in the British Isles tides are normally sinusoidal, but the precise shape of ebb and flow may be locally modified.

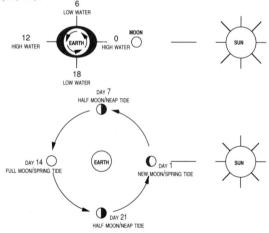

Fig. 1c. Spring tides occur when the gravitational pull of the sun and moon complement each other. This is maximal both at new moon (Earth, Sun and Moon in line) and at full moon (Sun and Moon opposite, Earth in between). Neap tides are caused when the Sun and Moon are at right angles to the Earth (rising and falling half moon). (*Modified from Fish & Fish, 1989*).

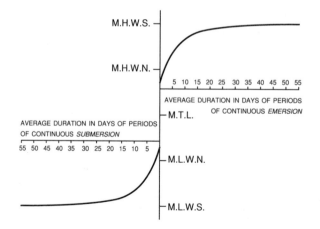

Fig 1d. Theoretical average duration of periods of continuous submersion or emersion at different levels on the shore. (*After Lewis, 1964*).

The seashore is the boundary between the land and the sea. A very sharp change in environmental conditions occurs between the fully marine habitat below the low tide mark and the fully terrestrial environment above the influence of sea-spray (Fig. 1a). This gradient of conditions results from the degree of wetting caused by the twice daily **ebb** and **flood** of the tide and by wave action (Fig. 1b). It is an essentially *vertical* gradient. **Spring tides** are of large amplitude (range) and occur when the gravitational pull of the moon and sun act in concert on the earth's water mass (Fig. 1c). **Neap tides** are of smaller amplitude and occur when the actions of the moon and sun are opposed. There is a regular two-weekly cycle of neap to spring to neap tides. In the spring and autumn, spring tides are at their largest amplitude, rising and falling the furthest. Fig. 2 lists various terms used to help describe tidal levels.

At increasing levels above the low water mark, the shore will be covered with water for shorter periods of time. At levels above the high-water mark for neap tides, the shore can go for several days without immersion, unless stormy conditions raise the water level or cause splash and spray (see Fig. 1d). The amplitude and timing of the tides vary round the British Isles. Tidal ranges less than 2-3 m are found around the Solent and the Firth of Clyde, while the largest tides (up to 10 m or so) are found in the Bristol Channel, Channel Islands and the eastern Irish sea (see Lewis, 1964 for details). **Double low and high tides** can occur from Portland to the Solent. Admiralty tide tables produced each year contain details of times and ranges of tides and shapes of the tide curve.

Most shore plants and animals probably evolved from marine ancestors. Very few species have invaded the shore from the land. Consequently, for most shore species a uni-directional **stress gradient** exists due to the increased proportion of time exposed to the air from low to high levels on the shore. The sea, even in inshore waters, is a remarkably stable environment. The **salinity** (concentration of dissolved salts or ionic composition) is virtually constant and temperature changes are small and slow. Carbon dioxide and (obviously!) water needed for plant photosynthesis are abundant. The nutrients such as nitrates and phosphates needed for protein synthesis and hence growth are also readily available, and rarely limiting in coastal waters. Therefore, when exposed to air, rocky shore plants and animals are subject to much greater stress and fluctuations in environmental conditions than when submerged. Air temperatures fluctuate rapidly; the air may be dry or it may rain; daily and seasonal extremes are also much more pronounced. When the tide recedes, sunlight, warm temperatures and dry air can all combine to lower the relative humidity on the shore and cause problems of desiccation for the plants and animals. In rock pools, salinities can be reduced by rainfall and be increased by evaporation causing **osmotic** problems. Very high light levels are thought to be damaging to some algae particularly delicate reds. These physical and chemical stresses are greater, more frequent and more variable the higher up in the intertidal a particular organism lives. Most of the above effects are chronic and sub-lethal. Occasional extreme or unexpected hot weather or frosts will kill intertidal life. Acute events affecting the mid and lower shore are uncommon in the British Isles, except for damage to low shore kelps and red algae during spring and summer in some years. The small **ephemeral algae** of the upper shore, however, get killed every spring, except in the far north.

There are also biological problems which increase at higher tide levels. Algae must take oxygen, water, carbon dioxide and other nutrients in through their fronds (**thalli**) in order to respire, photosynthesise and grow. Unlike terrestrial higher plants they do not have roots, so they need to be surrounded by water. Many intertidal animals respire more effectively in water than air because of gill-like gas exchange structures. However, there are many adaptations allowing intertidal animals to respire in air. Limpets trap a thin film of water under their shell and, by slightly lifting the shell, air diffuses into this water and so into their modified gills. Alternatively, some animals, such as mussels, can trap water within the body cavity and respire **anaerobically**, building up an oxygen debt which is paid-off by **aerobic** respiration on the next incoming tide. Animals which filter their food from the water (e.g. barnacles, mussels, tubeworms) or catch small prey items carried past them (e.g. sea anemones and hydroids) can feed only when immersed. Many mobile animals, including limpets, winkles and dogwhelks, move and feed (**forage**) most effectively when the tide is in, or during low tide periods with favourable damp conditions, such as at night or on humid days. Predation pressure from larger mobile animals, such as fish or crabs, which migrate into the intertidal from offshore, will be greater lower on the shore. Predators permanently resident on the shore, will also have a longer submerged feeding time lower down. Conversely predation by birds will be greater at higher levels. This is often overlooked as biologists scare birds away when working!

The second major environmental gradient is caused by different degrees of wave action (usually termed **exposure**) between **sheltered** bays and **exposed** headlands. It is essentially a *horizontal* gradient. Stress does not occur in a clearly defined direction on this exposure gradient. Problems of dislodgement increase with wave action and foraging may become difficult for some mobile animals. Conversely, in sheltered conditions, deposition of silt may clog gills and smother plants and animals. Nutrient uptake may be increased in plants with some water movement breaking down **boundary layers** near to the thallus and, similarly, **filter-feeders** will have more suspended food particles brought to them in more exposed areas. Desiccation problems, particularly higher on the shore, may be lessened due to wave spray on exposed shores. This allows marine organisms to live higher up on exposed rock surfaces than on sheltered ones.

The major vertical and horizontal environmental gradients are modified by a whole suite of factors primarily generated by the geology and topography of the shore. The aspect of the shore, or a particular rock on the shore, will modify the degree of desiccation or exposure. Generally, species will extend higher on the shore in the shade on north faces. Very few shores are unbroken 45° slopes, they are criss-crossed by crevices, fissures and gulleys, pock-marked by rock pools and have overhangs and raised areas caused by rock outcrops and boulders. The biota can also modify environmental

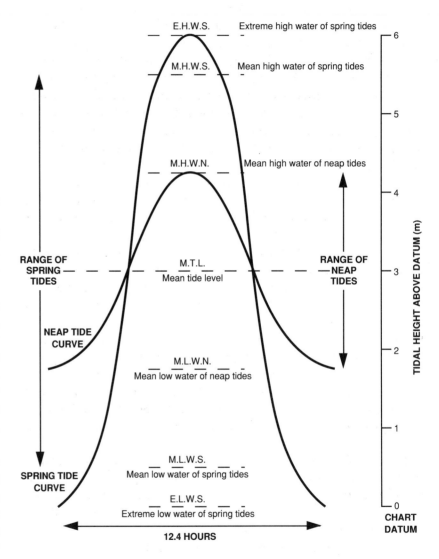

Fig. 2. Terms used in describing tidal levels and the shape of typical spring and neap tide curves.

conditions locally; many species will shelter under clumps of seaweed or mussels, for example. The specific environmental differences in any one place on the shore will depend on these **microhabitat** characteristics. Is it a sheet of well-drained rock; is it a pool; is it a crevice; is it a shaded overhang; is it north- or south-facing? The nature of the rock itself is also important; soft crumbling sandstone, although retaining moisture, does not offer seaweed holdfasts or barnacle shells such a secure anchorage as a harder rock like limestone or granite. Obviously the type of rock and its position affects the degree of **weathering** and the variety of habitats.

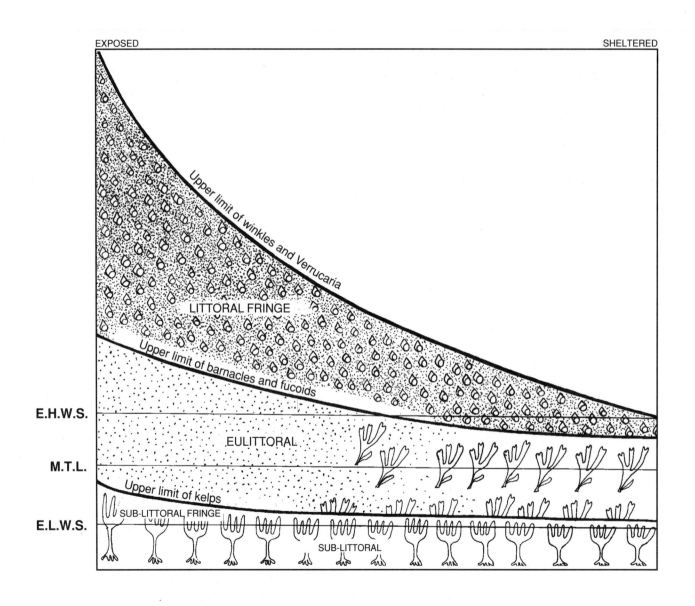

Fig. 3. A diagram illustrating the terminology applied to the main zones of rocky shores. (*Modified from Lewis, 1964*).

Stephensons' 3-Zone System

In the 1930s and 1940s, a husband and wife team of biologists, the Stephensons, travelled the world describing the communities on rocky shores. In 1946, they proposed a "universal" scheme of zonation for all rocky shores. They split the shore into three major zones: a high-shore zone, characterised by lichens and small winkles; a broad mid-shore zone, dominated by filter-feeding barnacles or mussels and a more narrow low-shore zone dominated by red algae, kelps or in some places in the southern hemisphere large filter-feeding tunicates (sea-squirts). This bottom zone is essentially an extension of fully marine conditions below the low tide mark and is exposed mainly during spring tides. The Stephensons based their classification on the communities present and were at pains to point out that this biological classification did not correlate with any set tide levels or physical conditions (see Fig. 2). They also stated that the scheme was most applicable to the shores of moderate, but not extreme, wave action.

In their 1972 book (completed after T. A. Stephenson's death) summarising their studies, some qualifications were added and "universal" was toned down to "widespread". Despite some detractors, the scheme does seem to apply to most shores around the world, with objective quantitative methods (e.g. Russell, 1973) also verifying the Stephensons' more subjective, intuitive classification. Therefore, the 3-zone system forms a useful descriptive framework for studies of the shore. In this book, we have used a slight modification of Lewis's (1964) version of the Stephensons' terminology (see Fig. 3): the top zone is called the **littoral fringe**, the middle zone the **eulittoral** and the low-shore zone the **sublittoral fringe** which is taken as being that part of the **sublittoral zone** which is exposed at low water.

In the north-east Atlantic the scheme works reasonably well. The exception is on sheltered shores where it is difficult to draw a meaningful biological distinction between an area covered by a kelp canopy or a wrack canopy. This is not really surprising, as the Stephensons only wished to apply their classification to shores with some wave action. Most authorities, however, would distinguish the sublittoral fringe by the presence of kelps and delimit the eulittoral on the presence of fucoids (wracks), thus the classification can be extended to sheltered shores (see Fig. 3). It is important to remember that these zones are biologically defined – that is by the organisms present. In many cases boundaries may be sharp – but some boundaries may be less clear.

The identification section of this book details the species used to indicate the littoral fringe, eulittoral and sublittoral fringe. The littoral fringe commonly has a characteristic black colour due to sheets of the lichen *Verrucaria maura* and blue-green bacteria (**cyanobacteria**) covering most of the rock surface. The tiny black winkle, *Melaraphe neritoides*, (previously called *Littorina neritoides*), is common on moderately exposed to exposed shores. The **polymorphic** multi-coloured winkle *Littorina saxatilis* is also common in this zone. Both winkles favour crevices, especially during dry weather. The upper limit of where barnacles are found in quantity delimits the boundary between the littoral fringe and eulittoral in all but sheltered conditions. A few scattered barnacles in pits, pools or crevices can be found in what is strictly the littoral fringe. In more sheltered conditions, the top of the *Pelvetia* zone, the uppermost band of fucoids (wracks), signifies the littoral fringe/eulittoral boundary. At exposed sites, the eulittoral will be dominated by barnacles (*Chthamalus montagui*, *C. stellatus* or *Semibalanus balanoides*, depending on geography) and sometimes mussels (*Mytilus* spp.). Shores of intermediate exposure usually have patches of fucoids which consolidate to cover the whole of the eulittoral as more sheltered conditions are approached. The boundary between the eulittoral and the sublittoral fringe is often less sharp. The upper limit of laminarians (kelps) is the best guide, although many species of red algae will straddle this demarcation and sometimes blur it.

Despite the Stephensons' protests to the contrary, the three zones do broadly correspond with different physical environments. The littoral fringe is rarely submerged except by spring tides or during the heaviest storms, although it maybe subject to continual splash or spray. The eulittoral is usually covered and uncovered by two tides every day. The sublittoral fringe is only exposed during spring low tides or on calm days. The three zones respond to differences in conditions along the wave gradient. In exposed conditions, upward wetting by salt spray can expand and elevate the littoral fringe to a height of tens of metres. On sheltered shores, it is very narrow and often bounded by terrestrial vegetation colonizing downwards. Similarly, the upward limits of the eulittoral and, less markedly, the sublittoral fringe will be elevated in exposed conditions.

Zonation

Using the Stephensons' 3-zone system as a broad framework, we can fill in the details of distribution patterns on British shores. Fig. 4 is a generalized diagram for British west-coast shores showing changes from shelter to exposure, modified from Ballantine (1961). Combined with the more detailed view of the three types of shore given in Fig. 5 and Plate I, a "typical" pattern of zonation can be described. There are some local variations to this general theme and for a fuller picture the reader is recommended to Lewis's (1964) excellent account of distribution patterns on a multitude of shores throughout the British Isles.

Sheltered shores

The most conspicuous feature of sheltered shores is the dense growth of large seaweeds in the eulittoral (Plate I, 1). These fucoids (wracks) occur in distinct zones, giving way to the laminarians (kelps or oarweeds) lower down in the sublittoral fringe.

Above the eulittoral is a very narrow lichen/winkle dominated littoral fringe (Plate I, 2). The encrusting lichen *Verrucaria* and bluegreens (cyanobacteria), which give the rock a distinctive black colour, dominate this zone. Many *Littorina saxatilis* may be present.

The following fucoid zones occur in descending order in the eulittoral: *Pelvetia canaliculata, Fucus spiralis, Ascophyllum nodosum* and *Fucus serratus*. Sometimes, (e.g. on the Isle of Man) *Fucus*

vesiculosus occurs in a narrow zone above *Ascophyllum*. Usually *F. vesiculosus* occurs in isolated clumps throughout the *Ascophyllum* zone, and is sometimes found below the *Ascophyllum* zone mingling with *F. serratus*. It is outcompeted by *Ascophyllum*, but it is highly opportunistic, colonizing clearings amongst the *Ascophyllum* faster than the *Ascophyllum* itself can recolonize, hence the occasional clumps. *Ascophyllum* rapidly decreases in cover as wave action increases and on a moderately sheltered shore *Fucus vesiculosus* completely replaces *Ascophyllum* and can form an almost complete cover on the mid-shore.

Barnacles and limpets are present in low numbers under the fucoid canopy on sheltered shores. If anything, they are more common above the *Ascophyllum* zone amongst *F. spiralis* and *F. vesiculosus*. In addition, many *Littorina saxatilis* occur in the *Pelvetia* and *Fucus spiralis* zones. Amongst the *Ascophyllum, Littorina obtusata* is abundant and further down the shore amongst the *Fucus serratus, L. mariae* is most common. *Littorina littorea* are often found on sheltered shores, but their distribution is patchy, being rare in some localities, presumably due to vagaries of larval dispersal. *Gibbula umbilicalis* can be common under *Ascophyllum*, whereas *Nucella* is rare on very sheltered shores, where, in the absence of barnacles and mussels they feed on littorinids.

Under the *Ascophyllum* canopy, a dense underturf of algae occurs; many species are present forming an interlocking matrix which often traps silt. *Audouinella* (formerly *Rhodochorton*) traps sand in particular, and here *Cladophora, Corallina, Chondrus, Lomentaria, Mastocarpus* (formerly *Gigartina*) and *Laurencia* are all common, together with many other species not listed in this book. Rock kept free by limpet grazing may be covered by "lithothamnia". Limpets often reappear in the *Fucus serratus* zone, the rock here being covered by "lithothamnia" with patches of *Cladophora, Corallina, Chondrus, Mastocarpus* and *Palmaria*, plus some encrusting sponges, bryozoans and hydroids. A similar understorey is found under the laminarians.

In the sublittoral fringe on very sheltered shores, *Laminaria*

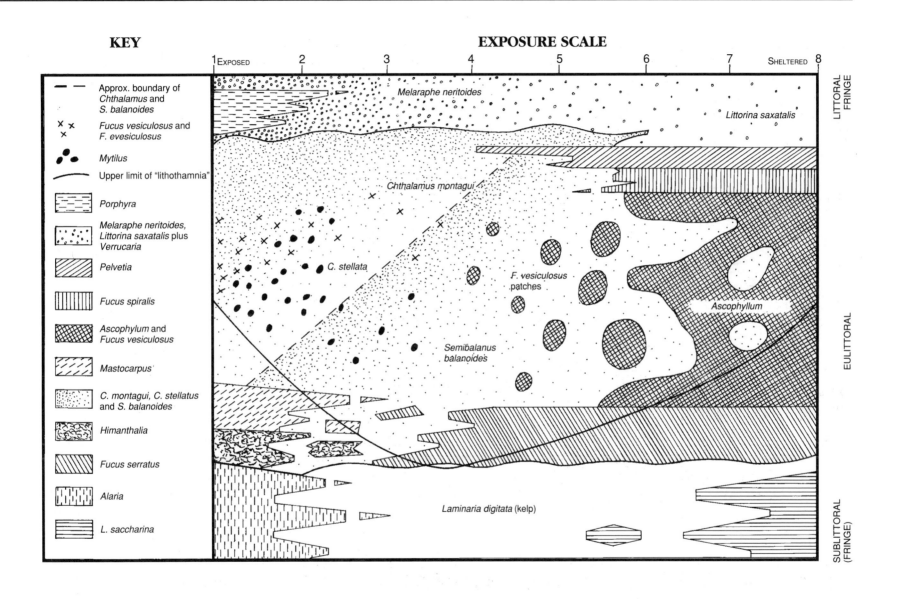

Fig. 4. Generalised diagram illustrating the changes in distribution of shore organisms on a west-coast shore in a gradient from sheltered (graded 8) to exposed (graded 1). For ease of presentation the vertical extent of each zone is compressed to constant height. (*Modified from Ballantine, 1961*).

PATTERNS OF DISTRIBUTION ON ROCKY SHORES

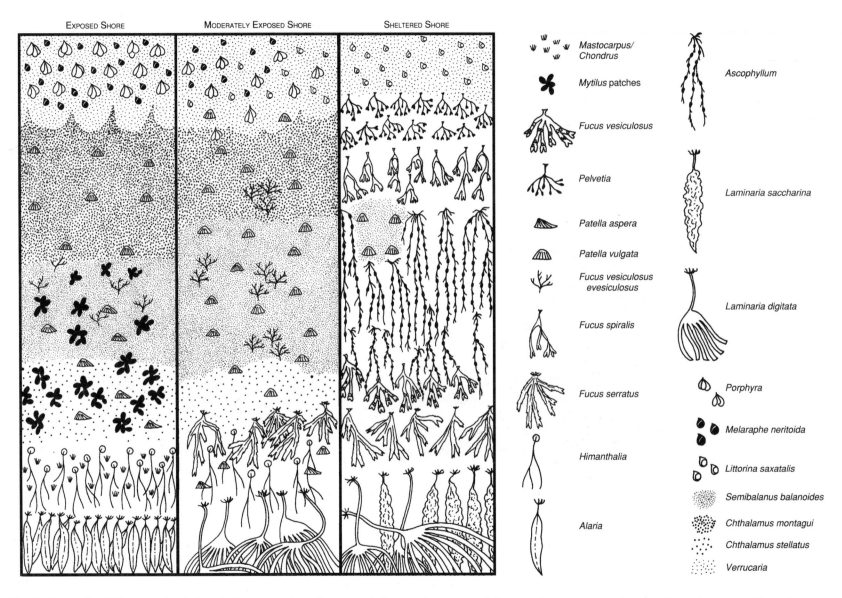

Fig. 5. Generalised diagram of a sheltered shore, a moderately exposed shore and an exposed shore on the west coast, showing the main zones of important shore organisms. (*After Ballantine, 1961*). Compare with Plate 1.

saccharina occurs beneath the *Fucus serratus* zone, often on boulders. *Laminaria digitata* gradually replaces *L. saccharina* as exposure increases, especially on bedrock.

Moderately exposed shores

With increasing wave action the littoral fringe broadens; in the eulittoral, fucoid cover begins to break up into patches forming a mosaic of barnacles, bare rock, *Fucus* and limpets; in the sublittoral fringe a dense band of kelp, *Laminaria digitata,* marks the bottom of the shore. Above the kelp there is often an algal turf zone of various species.

The broader littoral fringe is still dominated by *Verrucaria maura*, with *Littorina saxatilis* and some *Melaraphe neritoides*. In winter and spring, dense growths of *Prasiola, Blidingia, Porphyra, Enteromorpha* and green, filamentous *Ulothrix* can occur. The rock often has a black or brown stain due to blue-greens (cyanobacteria) and diatoms.

As wave action increases, fucoid cover is reduced further. This is most marked in the mid-eulittoral. High in the eulittoral, *Fucus spiralis* can form a dense canopy on quite exposed shores and *F. serratus* will also extend into quite exposed conditions. Any remaining *Ascophyllum* plants are exceedingly stunted (Plate III, 4). The mid-shore becomes a mosaic of patches of *Fucus vesiculosus*, barnacles, bare rock and limpets (Plate I, 3 & 4). *F. vesiculosus* becomes increasingly stunted and the bladderless morph (*Fucus vesiculosus evesiculosus*) more dominant, although some bladdered plants are still to be found. During the spring, patches of ephemeral green algae (mainly *Enteromorpha* spp.) are often found on the mid-shore, as well as occurring in pools and damp places in the littoral fringe. Patches of *Laurencia* occur in the lower eulittoral and on shaded rocks. Steeply sloping shores tend to be barnacle-dominated. *Chthamalus montagui* occurs in the mid and high eulittoral in the south-west with a mixture of *Semibalanus balanoides* and some *Chthamalus stellatus* lower down. Further north, only *Semibalanus* will be found over much of the shore and *C. montagui* becomes increasingly restricted to a narrow high shore band. Any *C. stellatus* will be restricted to the overlap zone between *S. balanoides* and *C. montagui*. On the east coast of Britain only *S.balanoides* will be found. *Lichina pygmaea* is characteristic of barnacle-covered, sunny, steep rocks in the upper eulittoral.

Throughout the eulittoral, large numbers of *Patella vulgata* occur on all coasts. In the south and south-west (except Ireland) up to 50% of the population may be *Patella depressa*; they are especially common from mid-tide level (MTL) upwards. A few *Patella aspera* will be present on open rock low on the shore and in shallow "lithothamnia"/ *Corallina* rock pools at all levels. *Gibbula umbilicalis* and *Littorina littorea* occur at mid-shore level on most moderately exposed shores, joined by *Monodonta lineata* in the south and west. *Gibbula cineraria* is found mainly in pools and under stones on the lower shore and can be quite common. Dogwhelks are common in clumps of seaweed or in crevices, feeding primarily on barnacles, although they will take small limpets. Mussels are sometimes found on moderately exposed shores, but this is not the general rule. Amongst the barnacles, small littorinids thrive (*L. nigrolineata* and *L. neglecta*). *Littorina obtusata* and *L. mariae* do occur, but with the reduced abundance of fucoids, their habitat plants, they become rarer.

The anemone *Actinia* is common in pools, crevices under fucoid clumps, and on open rock on moderately exposed shores. Another anemone *Anemonia*, only occurs in standing water pools.

Below *Fucus serratus*, at the boundary of the eulittoral and the sublittoral fringe, dense turfs of algae occur. *Chondrus, Corallina, Cladophora* , *Mastocarpus* and *Palmaria* are all common, with the rock often taking on a characteristic pink colouration due to the "lithothamnia". The thong-weed, *Himanthalia elongata*, is present both as the immature buttons and as the strap-like mature plants. The sublittoral fringe is dominated by *Laminaria digitata* at lower levels. Occasional *Alaria* plants occur, except on western English Channel shores. On the lowest tides, the erect stipes of *Laminaria hyperborea* become exposed.

In the sublittoral fringe, the painted topshell, *Calliostoma zizyphinum* occurs. The blue-rayed limpet, *Helcion pellucidum* (sometimes called *Patina pellucida*), is common on kelp stipes and also under the holdfasts. The sponges, *Hymeniacidon* and *Halichondria*, plus *Corallina* are often common under the kelp canopy.

Exposed shores

The most noticeable feature of exposed shores is the very broad *Verrucaria*-covered littoral fringe, which can extend 30-40 metres up cliffs on very exposed headlands (Plate I, 5). *Melaraphe neritoides* is extremely abundant in pits and crevices, where *Littorina saxatilis* can also be very common but lower down. *Porphyra* is usually present in the lower littoral fringe forming a dense band in the winter or spring.

The eulittoral is usually covered by barnacles or mussels. Amongst the latter, a scattering of the small bladderless *Fucus vesiculosus evesiculosus* (or form *linearis*) may occur, as well as some small red algae, primarily *Ceramium* spp. The species of barnacle depends on geography. In the South and West, *Balanus perforatus* can form a narrow zone low in the eulittoral extending into moderate shelter. *Chthamalus montagui* occurs high in the eulittoral on all exposed western coasts, whilst at mid to low shore levels in the south-west *Chthamalus stellatus* is the dominant barnacle. *Semibalanus balanoides* only extends into very exposed conditions in quantity on more northerly and easterly shores; in the far south it is mainly confined to moderate shelter. *Nucella* is abundant if there are suitable crevices for protection against wave action and limpets are exceedingly common everywhere. In south-west Britain, *Patella depressa* can be very common amongst mussels, forming more than 50% of the population. *Patella aspera* is found in large numbers on open rock, as well as in pools and, in extremely exposed conditions will extend above mid-tide level. On the Irish coast, the sea urchin *Paracentrotus* is common in pools in more exposed conditions, although it will also occur sublittorally in shelter (e.g. Lough Hyne). Many *Actinia* occur in gulleys or surge channels. In some places,

Gibbula umbilicalis and *Littorina littorea* can be very common on exposed shores, usually sheltering in gulleys. *Monodonta* are also often found in localized shelter high in the eulittoral. *Littorina neglecta* and *L. nigrolineata* are abundant amongst the barnacles. On the most exposed and steepest shores, the eulittoral has a very low diversity fauna, with only barnacles, mussels and limpets occurring on open rock.

The sublittoral is characterized by "lithothamnia" and algal turfs (Plate I, 6). *Chondrus*, *Mastocarpus* and *Corallina* are the main species. *Alaria* replaces *Laminaria digitata*, except in the English Channel, where it is absent. In the far south and west, a zone of *Bifurcaria* or *Cystoseira* can occur above the kelps; these species are also common in midshore pools.

Geographic variation

The overall pattern of zonation and community composition described above is relatively uniform throughout the British Isles. Some species do, however, show considerable geographic differences in their distribution. This is because the British Isles are at a boundary between an essentially warm-water, southern fauna and flora extending up from Spain and Portugal and **boreal**, cold-water species extending down from Norway. The southern species become rarer or disappear in the north and east. For example, the barnacles *Chthamalus montagui* and *C. stellatus* become increasingly restricted to the high shore as *Semibalanus* increases in dominance on more northerly west coast shores and in the eastern part of the English Channel. They are absent from east coasts. *Patella depressa* forms a decreasing proportion of the limpet population from St David's Head northwards and east of Swanage before reaching its respective northern and eastern limits on Anglesey and the Isle of Wight. The eastern limits of many southwestern species are between Plymouth and the Isle of Wight. Northern limits on the west coast are more diffuse with species dropping out as one proceeds clockwise around the British Isles. Some northern species, such as *Fucus distichus* and *Strongylocentrotus drobachiensis* begin to appear in northern Scotland. Fig. 6 summarises the geographic distribution of selected species around the British Isles. Generally, fucoids become more predominant and extend further into exposure on more northerly shores. To put the British Isles into perspective, Fig. 7 shows the distribution patterns of major species with wave action and tidal height in northern Spain, Wales and in Norway. Shores in south-west Britain, Ireland and the Channel Islands have similarities to the pattern found further south, whereas Scottish shores are, in many ways, similar to Norway.

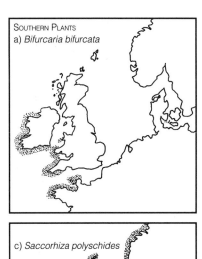

SOUTHERN PLANTS
a) *Bifurcaria bifurcata*

b) *Cystoseira* spp.

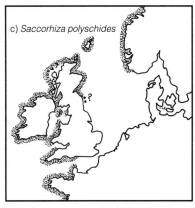

c) *Saccorhiza polyschides*

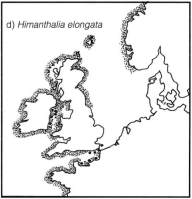

d) *Himanthalia elongata*

AN INVADER
e) *Sargassum muticum*

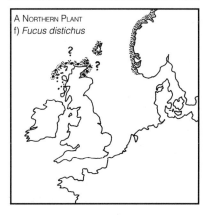

A NORTHERN PLANT
f) *Fucus distichus*

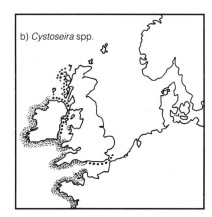

Fig. 6. Geographical distribution around the British Isles of selected intertidal species.
    ?   indicates uncertain status
    …   indicates previously common but much reduced due to overcollecting.

Fig. 6 continued

SOUTHERN ANIMALS
g) *Anemonia sulcata*

h) *Paracentrotus lividus*

i) *Monodonta lineata*

j) *Gibbula pennanti*

k) *Gibbula umbilicalis*

l) *Patella depressa*

m) *Patella aspera*

n) *Balanus perforatus*

o) *Chthalamus stellatus*

p) *Chthalamus montagui*

q) *Sabellaria alveolata*

NORTHERN ANIMALS
r) *Semibalanus balanoides*

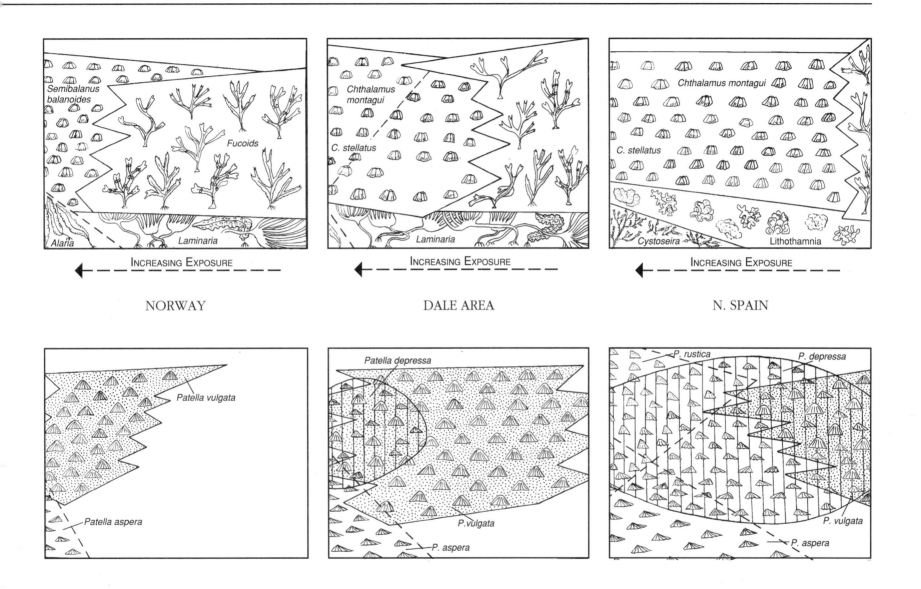

Fig. 7. Distribution patterns of main intertidal species in northern Spain, Wales and Norway to illustrate the effects of latitude. Upper diagrams: fucoids and barnacles; lower diagrams: species of *Patella*. (*After Ballantine, 1961*).

LIFE HISTORIES OF SHORE ORGANISMS

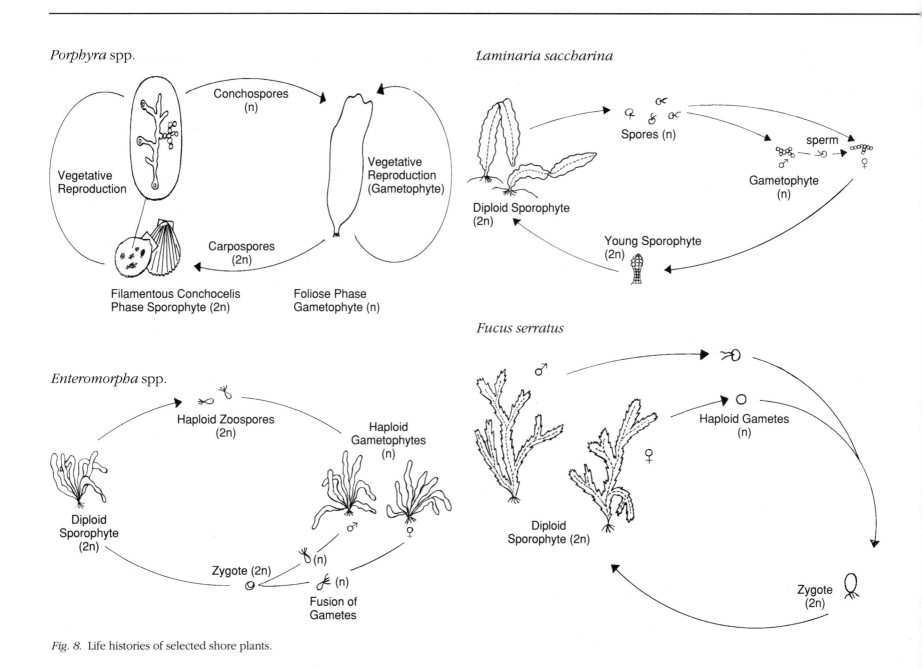

Fig. 8. Life histories of selected shore plants.

Algae

The life histories of the common algae on the shore are complex and varied, so the reader is referred to more advanced texts (e.g. Boney,1966; Dring,1982). Basically, there is an alternation of a **haploid**, **gamete**-producing phase (**gametophyte**-producing eggs and sperm) and a **diploid** spore-producing (**sporophyte**) phase. Single-phase life histories with the adult diploid plant producing gametes do occur (e.g. *Fucus*). All have dispersive stages which are washed around in the water before settling on the rock and growing into a tiny plant, or germling. Some spores are able to swim (e.g. *Enteromorpha*). Other plants such as *Ascophyllum*, many turf-forming reds and encrusting algae, can proliferate vegetatively as well as sexually. Some typical life histories are outlined in Fig. 8. Beware that even in the same genus different patterns can occur! *Fucus vesiculosus* and *F. serratus* have plants of separate sex, whereas *F. spiralis* has both male and female parts on the same plants. Some algae have alternate phases in their life histories which are so morphologically different that they were once thought to be different species (e.g. various crusts have erect phases: *Ralfsia/ Petalonia*, *Petrocelis/Mastocarpus* and *Porphyra* has an alternate burrowing 'Conchocoelis' phase

Animals

Fig. 9 outlines the life histories of selected shore animals. Many marine animals liberate eggs and sperm into the water for external fertilization, followed by a larval **planktonic** stage, which may last from hours to weeks. All the limpets, mussels, starfish and sea urchins listed in this book reproduce in this way. Sea anemones can reproduce either asexually by budding fission (splitting) or sexually leading to a larval stage which may be brooded. Hydroids have alternate generations: there is usually a medusa (jellyfish) phase in the plankton and the polyp attached-phase on the shore, although the medusoid stage has been lost in some species. Amongst the gastropods, external fertilization occurs in *Patella*, *Gibbula* spp. and in *Monodonta*, leading to planktonic trochophore larvae and thence to settling veliger larvae. In contrast, internal fertilization occurs in all species of winkle (littorinid). *Littorina littorea* and *Melaraphe neritoides* release their eggs into the plankton; *L. obtusata* and *L. mariae* lay their eggs in capsules on seaweeds while in the case of *L. saxatilis*, *L. nigrolineata* and *L. neglecta* eggs are brooded inside the female and hatch as live young. *Littorina arcana*, however, which is impossible to segregate from *L. saxatilis* in the field, lays its eggs on the rock. Dogwhelks have internal fertilization followed by the laying of egg capsules. *Patella vulgata* (and perhaps *Patella aspera*) change sex. When they first become sexually mature most are males, but as they become older some of the males become females, whilst others probably remain male all their lives. In consequence the larger older limpets are mainly female. See Graham (1988) for further details of reproductive biology of gastropods.

Internal fertilization occurs in all the Crustacea, followed by a brooding of eggs before release of larvae to the plankton. Even barnacles fertilize internally! They overcome the problem of being sessile by being cross-fertilizing **hermaphrodites** with a very long penis (2-3 times body length) which they insert into neighbours. Self-fertilization may also occur rarely.

Animals with larvae have greater dispersive ability than those without; this allows exploitation of new space or transiently available areas. Most larvae have behaviour patterns to ensure they settle in favourable conditions: barnacle cyprids prefer pits, rough areas, align with water currents and settle near adults of the same species; spirorbids are attracted to particular algal hosts; and mussels like fibrous textures. However, the production of crawling, live young from egg capsules or brood pouches cuts down on reproductive waste and allows exploitation of locally favourable conditions. It can also lead to inbreeding and genetic isolation of populations in isolated crevices or on boulders. This allows recessive characters to be expressed with resulting variation in shell colour and morphology.

LIFE HISTORIES OF SHORE ORGANISMS

BARNACLES: *Semibalanus balanoides*

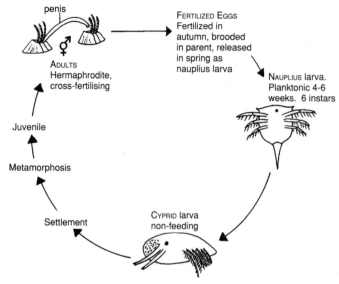

penis

ADULTS
Hermaphrodite,
cross-fertilising

Juvenile

Metamorphosis

Settlement

FERTILIZED EGGS
Fertilized in
autumn, brooded
in parent, released
in spring as
nauplius larva

NAUPLIUS larva.
Planktonic 4-6
weeks. 6 instars

CYPRID larva
non-feeding

Littorina littorea

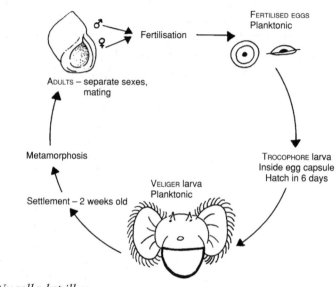

ADULTS – separate sexes,
mating

Metamorphosis

Settlement – 2 weeks old

FERTILISED EGGS
Planktonic

Fertilisation

TROCOPHORE larva
Inside egg capsule
Hatch in 6 days

VELIGER larva
Planktonic

LIMPETS: *Patella vulgata*

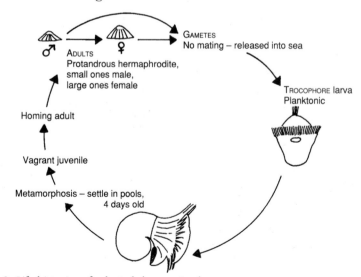

ADULTS
Protandrous hermaphrodite,
small ones male,
large ones female

Homing adult

Vagrant juvenile

Metamorphosis – settle in pools,
4 days old

GAMETES
No mating – released into sea

TROCOPHORE larva
Planktonic

Nucella lapillus

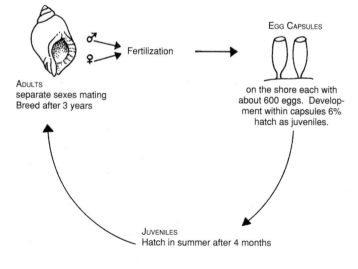

ADULTS
separate sexes mating
Breed after 3 years

Fertilization

EGG CAPSULES

on the shore each with
about 600 eggs. Develop-
ment within capsules 6%
hatch as juveniles.

JUVENILES
Hatch in summer after 4 months

Fig. 9. Life histories of selected shore animals.

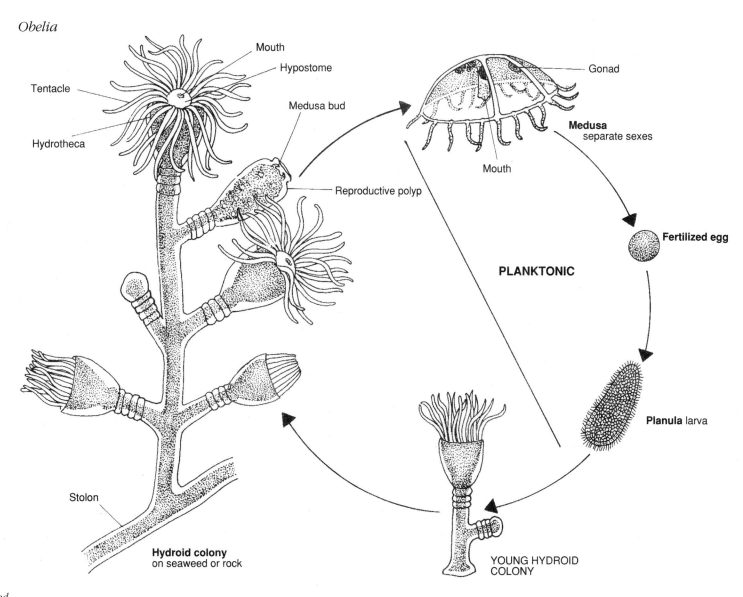

Obelia

Mouth

Hypostome

Tentacle

Medusa bud

Hydrotheca

Reproductive polyp

Gonad

Medusa separate sexes

Mouth

Fertilized egg

PLANKTONIC

Planula larva

Stolon

Hydroid colony on seaweed or rock

YOUNG HYDROID COLONY

Fig. 9 continued

IDENTIFICATION SECTION

We have used broad taxonomic groups to arrange the common species likely to be encountered on the shore, with the exception of a mixture of prominent space-occupying animals and prominent grazers and predators.

In the identification section we have deliberately not used keys. A tabular approach has been used with identification features indicated on line drawings facing the tables. A block of Plates in the middle of the book shows each species considered. The Plate number is indicated for each species in the tables (e.g. II, 3). Lists of more detailed identification guides and keys are given in the bibliography. Common names have been given in ( ) and other scientific names in general usage in [ ]. We have tried to be as up to date as possible with the names; some such as *Melaraphe* and *Mastocarpus* may surprise some people.

Many students and amateur naturalists are very puzzled by scientific Latin names. To overcome this problem we have included a short section outlining the system used for naming organisms.

Putting a name to intertidal organisms – some examples using barnacles

Many common marine plants and animals have an English or common name, but many do not. For example "barnacle" covers many similar organisms. "Acorn barnacles" distinguishes the sessile (non-stalked) species from "goose barnacles" which have a stalk, but there are several kinds of each not further differentiated by common names.

However, all known organisms have been given a scientific name according to a set of rules and these names are internationally recognised by scientists, whereas common names are not. The only accurate way to consistently refer to a species (a type of organism) is to use its scientific name – common names change from place to place. Unfortunately these can be quite difficult to remember, although some may be descriptive based on shape or colour of the organism. Even more confusing is that sometimes the scientific name of an organism is changed – usually for a very good reason.

The full name of the commonest British shore barnacle is:

Semibalanus balanoides (Linnaeus)

Note that the first two words are written in italics (underlined if typed or if hand-written). This is to distinguish it as the scientific name. *Semibalanus* is the GENUS to which the SPECIES *balanoides* belongs. They are always used together for the species name. Note also the capital initial letter of the genus *Semibalanus,* but not of *balanoides* the specific name. Linneaus is the name of the person (the authority) who first published a description of the species. His name is here in brackets because the species is now in a different genus to the one he put it in. Linneaus is sometimes abbreviated to L. In a scientific text the authority is usually given at the first mention of a species, further mentions will only give the genus and species, thus:

Semibalanus balanoides

Alternatively a local checklist or a key taxonomic paper (e.g. Isle of Man Marine Fauna, Bruce *et al.*, 1963) may be cited to save quoting the authority for each species. We have followed the practice of most elementary texts in omitting the authorities.

Sometimes to save space, the genus may be reduced to its initial after its first mention, thus:

S. balanoides

Before 1976, *S. balanoides* was known as *Balanus balanoides*. Why the change? It was because Newman and Ross (1976) reviewed the barnacles and decided that the genus *Balanus* contains only those barnacles with a calcareous base (they leave a white scar when taken off the rock) and certain other features. But *balanoides* has a membranous base and thus in their opinion, cannot be a *Balanus*. They put *balanoides* into *Semibalanus*, a genus of barnacles with a membranous base, hence *Semibalanus balanoides*.

Another name change has taken place (coincidentally also in 1976) concerning other intertidal barnacles. Two species of *Chthamalus* are now recognised on British shores:

Chthamalus montagui Southward
Chthamalus stellatus (Poli)

Before 1976 only one species, *C. stellatus*, was recognised following Darwin's classic work on barnacles. Southward (1976) re-examined the chthamalid barnacles and showed that what had hitherto been considered as one variable species was in fact two species. One retained the name given to it by Poli, but the other required a new name – hence *C. montagui* (named after Montagu, an early naturalist).

References
Newman, W. A. & Ross, A. 1976. Revision of the balanomorph barnacles; including a catalog of species. *Memoirs of the San Diego Society of Natural History*, No. 9, 108pp

Southward, A. J. 1976. On the taxonomic status and distribution of *Chthamalus stellatus* (Cirripedia) in the north-east Atlantic region: with a key to the common intertidal barnacles of Britain. *Journal of the Marine Biological Association of the United Kingdom*, 56, 1007-1028.

Lichens

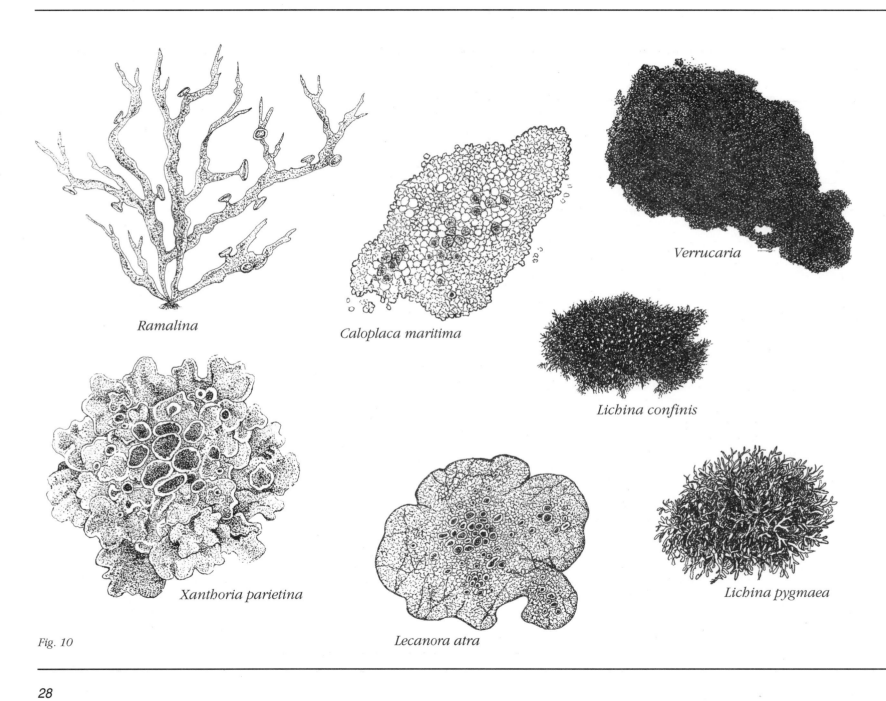

Ramalina

Caloplaca maritima

Verrucaria

Lichina confinis

Xanthoria parietina

Lecanora atra

Lichina pygmaea

Fig. 10

28

Lichens are closely encrusting or tufted plants, and are the dominant organisms in the littoral fringe and on the open rocks of cliffs above. They will also occur inland on any suitable bare hard structure. Some occur in the eulittoral.

Species	Distinguishing Features	Tidal Level	Exposure	Geography
Ramalina spp. II, 1	Pale grey/grey-green tufts. **Stiff upright**. 2-4 cm tall. Topped by convex discs.	Terrestrial down to upper limits of littoral fringe	Sheltered to very exposed	All coasts
Xanthoria parietina II, 1 & 2	Orange or yellow patches. Coarse thick **leavy** texture.	Terrestrial down to upper limits of littoral fringe	Sheltered to very exposed	All coasts
Caloplaca spp. II, 2	**Flat** encrusting **dark orange** patches with scattered granules. Not very leavy.	Terrestrial down to upper limits of littoral fringe	Sheltered to very exposed	All coasts
Lichina confinis II, 2	Small black patches, **finely tufted**. Almost velvet-like texture.	Upper half of littoral fringe	Sheltered to very exposed	All coasts
Lecanora atra II, 1	**Grey** irregular patches. **Central black** fruiting bodies.	Upper half of littoral fringe	Sheltered to very exposed	All coasts
Verrucaria spp.	**Oil-like** smooth black patches converging to give complete cover. **Not** tufted or "leaved" 2 species – not normally necessary to distinguish for surveys:	Littoral fringe to mid-eulittoral	Sheltered to very exposed	All coasts
V. maura II, 2 & 3; VIII, 2	**Not** green, extensive sheets.	**Dominant in littoral fringe**	Sheltered to very exposed	All coasts
V. mucosa II, 4	**Greenish**-black. patches < 50 cm.	Eulittoral	Moderately sheltered to exposed	W coasts
Lichina pygmaea II, 5	**Black, coarsely** tufted patches. 1 cm thick. 10-20 cm diameter. Resembles very dark red seaweed.	Upper eulittoral amongst barnacles	Moderately exposed to very exposed	All coasts

Identification notes: *Verrucaria* can sometimes be mistaken for an oil-stain, or the rock colour. Scratching the surface of the rock reveals that it is in fact covered by a fine sheet of material. It gives the upper shore its distinctive black colour, although this can be due to extensive sheets of blue-greens. [Further reference: B & Y: p. 252; C: p. 62; Dobson, 1981]. Other lichen-like plants are crust-forming brown and red algae plus calcareous "lithothamnia" (pp. 37, 39, 41).

Ecological notes: Lichens are composite organisms resulting from the symbiosis of an alga or a blue-green bacterium and a fungus. *Lichina pygmaea* clumps are a haven for many types of small invertebrates. Lichens are probably terrestrial in evolutionary affinity and have invaded the shore from the land. They are extremely slow growing and very susceptible to pollution.

Important Green Algae

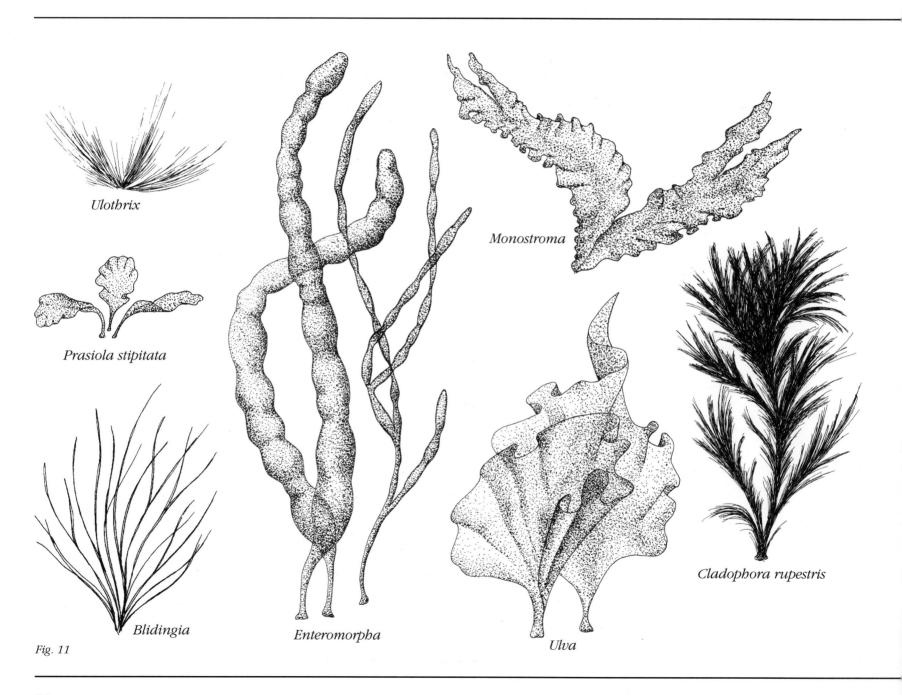

Ulothrix

Prasiola stipitata

Monostroma

Blidingia

Enteromorpha

Ulva

Cladophora rupestris

Fig. 11

There are numerous species of green algae (Chlorophyta) found on rocky shores. Those listed below are considered ecologically important. There are many varieties and species within the genera *Ulva* and *Enteromorpha* which cannot be identified without a microscope.

Species	Distinguishing Features	Tidal Level	Exposure	Geography
Ulothrix flacca II, 6	Single fine, light green threads in tufts **forms a fine sheet on rock** when tide out.	Upper littoral fringe (winter and spring)	Moderately exposed to exposed	All coasts
Prasiola stipitata II, 7	Very small, dark-green plant **Stalk is half total length**, which expands into an oblong, curled edge frond.	Littoral fringe (winter and spring)	Moderately exposed to exposed	All coasts
Blidingia minima [=*Enteromorpha minima*] II, 8	Small, soft, **grass**-green delicate strips which may be branched singly. Sometimes inflated (< 5 cm) **Forms short turfs high on shore**.	Littoral fringe and upper eulittoral	All exposures	All coasts
Enteromorpha spp. II, 9	**Frond tubular, or long and thin** (max. size 25 cm). Difficult to separate from *Blidingia* but tends to be a darker green.	All levels, mainly high and midshore	All exposures	All coasts
Ulva lactuca (Sea Lettuce) II, 10	Large, rounded, flat fronds (4-20 cm). **Firm when rolled between thumb and finger** as has 2 cell layers.	Lower eulittoral and in pools	Sheltered to moderately exposed	All coasts
Monostroma grevillei II, 11	Large, rounded flat fronds like *Ulva*, but **thin and slimy** when rolled between thumb and finger as 1 cell thick.	Lower eulittoral in pools	Sheltered to moderately exposed	All coasts
Cladophora rupestris II, 12	Forms short 3-8 cm turf, which can be extensive. Individual plants consist of **many dark-green, branching filaments.**	Lower eulittoral, higher under canopies	Sheltered to moderately exposed	All coasts

Identification notes: Separation of the various species of *Enteromorpha* and *Ulva* needs microscopic identification and they are best lumped in the field. *Ulva* and *Monostroma* can be lumped together but are easy to tell apart by touch. *Ulva* has a much stiffer frond. There are several species of *Cladophora*, all of which consist of fine branching filaments. Species other than *C. rupestris*, are generally paler green.

Ecological notes: With the exception of *Cladophora rupestris*, turfs of which persist for many years, all the above species are seasonal or short-lived (ephemeral) algae. *Prasiola stipitata* is restricted to the high shore; *Ulothrix*, *Blidingia* and *Enteromorpha* are more common high on the shore but occur opportunistically at all levels down when rocks overturn, if a new surface appears due to erosion, in grazer-free patches or on the shells of limpets. *Enteromorpha* is also associated with freshwater run-off. They are all early colonizers. Similarly, *Monostroma* and *Ulva* are low shore opportunists.

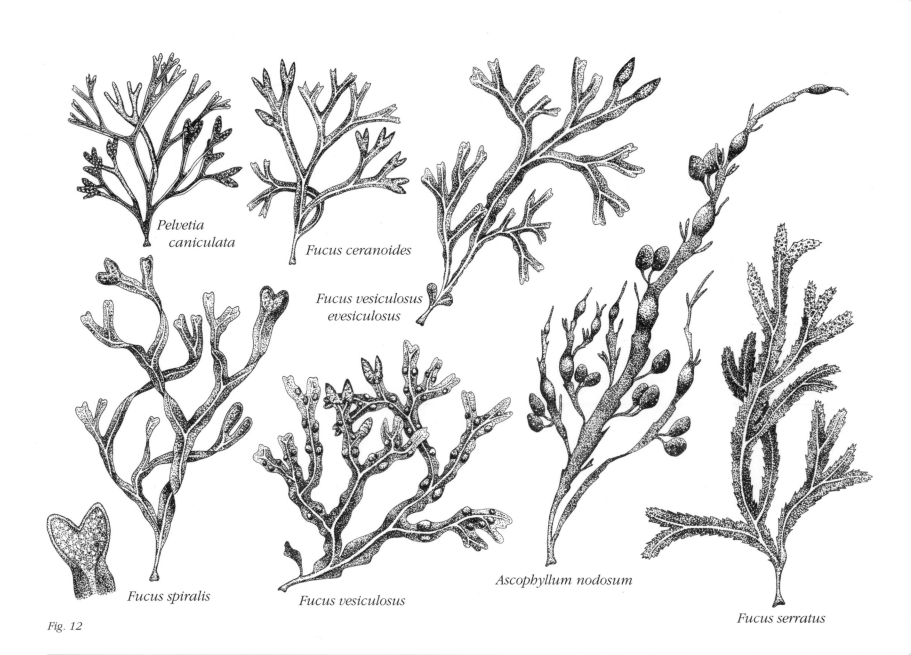

Pelvetia caniculata

Fucus ceranoides

Fucus vesiculosus evesiculosus

Fucus spiralis

Fucus vesiculosus

Ascophyllum nodosum

Fucus serratus

Fig. 12

The mid-shore regions of sheltered shores in the North Atlantic are dominated by large, brown seaweeds (Phaeophyta), the wracks or fucoids. There are several species which occur in characteristic horizontal bands or zones given in descending order below.

Species	Distinguishing Features	Tidal Level	Exposure	Geography
Pelvetia canaliculata (Channelled Wrack) II, 13	Small plant without bladders. **Fronds with channel.** (< 15 cm.)	Upper eulittoral	Mainly sheltered shores	All coasts
Fucus spiralis (Spiral Wrack) II, 14	Fronds twisted or spiralled. **Fruiting bodies have sterile margin.** No bladders. (30 cm.)	Upper eulittoral	Mainly sheltered shores	All coasts
Fucus ceranoides II, 15	Small plant without bladders but inflations in frond. Pronounced midrib. No channel. **Pointed fruiting bodies on frond tips.**	All levels where salinity is reduced	Estuaries and where freshwater run-off over shore	All coasts
Fucus vesiculosus (Bladder Wrack) III, 1	Very variable. Usually with **paired bladders.** Fruiting bodies **do not** have sterile margin.	Mid eulittoral	Commonest on moderately exposed shores	All coasts
F. vesiculosus evesiculosus [or *F. vesiculosis v. linearis*] III, 2	Bladderless form of *Fucus vesiculosus.* Separated from *F. spiralis* by lack of twisting and **no sterile margin of fruiting bodies.**	Mid eulittoral	Exposed and moderately exposed	All coasts
Ascophyllum nodosum (Knotted Wrack) III, 3 & 4; V, 3, 4 & 5	**Large single bladders.** A long plant (50-150 cm). Olive-green. Many fronds forming clumps. Often with red tufts of *Polysiphonia lanosa.*	Mid eulittoral	Sheltered shores occasional stunted bushes in moderate exposure	All coasts
Fucus serratus (Toothed Wrack) III, 5	No bladders and flat fruiting bodies. **Serrated edge to frond.**	Lower eulittoral	All exposures, commonest in shelter	All coasts

Identification notes: All fucoids are very variable, particularly *F. vesiculosus* which displays tremendous variation in numbers of bladders and plant size on gradients between shelter and exposure. On sheltered shores *F. vesiculosus* can grow up to 1 m with large numbers of paired bladders. As exposure increases it gets progressively more stunted with fewer bladders. In exposed conditions short bushy, erect plants (*F. v. evesiculosus*) grow which become sexually mature at 10 cm. Hybrids are also thought to occur and strange morphologies are common following disturbance. Another species, *F. distichus* occurs in the far north of Scotland to the arctic. Its taxonomy has just been revised and is beyond the scope of this book.

Ecological notes: Fig. 4 shows how the distribution of fucoids changes with wave exposure. On sheltered shores they form distinct horizontal bands, or zones, at different shore levels. *Ascophyllum* reproduces primarily from vegetative proliferation from existing clumps which are thought to be tens to hundreds of years old.

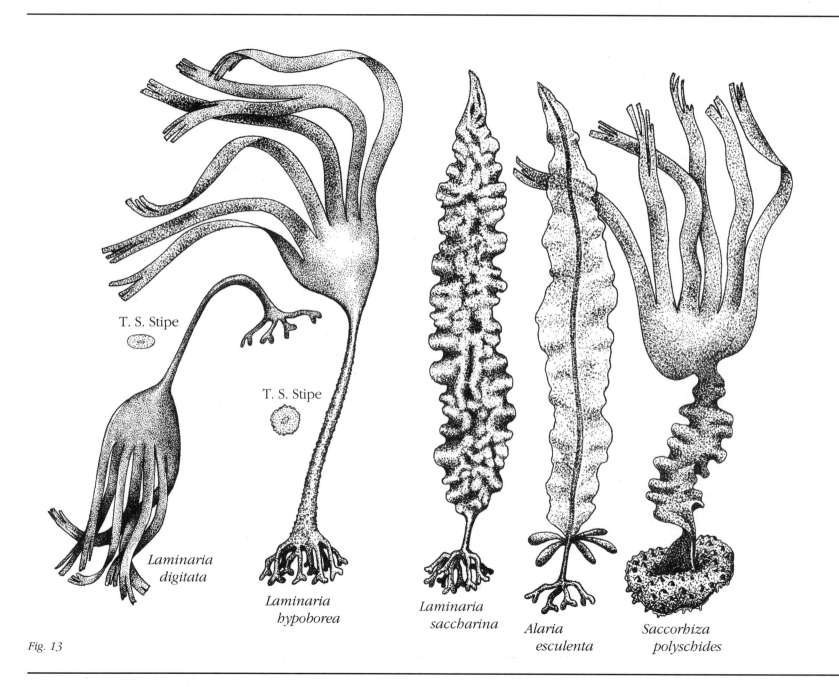

T. S. Stipe

T. S. Stipe

Laminaria digitata

Laminaria hypoborea

Laminaria saccharina

Alaria esculenta

Saccorhiza polyschides

Fig. 13

The sublittoral fringe is usually dominated by extensive forests of large brown algae (> 1 m). These extend into the subtidal to a depth of about 20 m.

Species	Distinguishing Features	Tidal Level	Exposure	Geography
Laminaria digitata (Oarweed, Tangleweed) III, 6 & 10	Flexible **smooth** stipe, **oval in cross-section**, particularly at join with frond. Lies **flat** on the shore when tide is out. Holdfast spread, divided frond. (< 2 m.)	Sublittoral fringe and lower eulittoral pools	Moderately sheltered to exposed, most common in moderate exposure	All coasts where firm, stable substrate
Laminaria saccharina (Sea Belt, Sugar Kelp) III, 8	Short stipe, long wavy **undivided** frond. (< 4 m.)	Sublittoral fringe and lower eulittoral pools	Sheltered (Can transiently occur in more exposed areas)	All coasts
Alaria esculenta III, 7	Short stipe, **frond with mid-rib**. Sometimes with small reproductive fronds attached to stipe. Frond often eroded. (< 2 m.)	Sublittoral fringe and lower eulittoral pools	Mostly exposed conditions	N, E and W coasts, Lizard to Yorkshire, absent W. English Channel
Laminaria hyperborea III, 9 & 10	Rigid **rough** stipe, **rounded in cross section**. Stands upright when tide is out. Holdfast more cone-shaped, divided frond. (< 3 m.)	Sublittoral fringe and below	Moderately sheltered to exposed, most common in moderate exposure	All coasts where firm, stable substrate
Saccorhiza polyschides [=*S. bulbosa*] III, 9	Large, knobbly, bulbous, hollow holdfast. **Flattened stipe**, often with wavy edges. (< 2 m.)	Sublittoral fringe and below	Opportunist, found in all exposures	More common in south and west, absent Solent to Northumberland

Identification notes: The young plants are difficult to identify. The stipe characteristics are best for this.

Ecological notes: *L. digitata* and *L. hyperborea* form dense canopies that reduce light penetration to the plants below. The fronds also exert considerable sweeping effects. Often the flora under these canopies is poor and specialized. *Alaria* is an opportunist species, which thrives in exposed or unstable conditions where the dominant canopy of *L. digitata* and *L. hyperborea* has broken down. *L. saccharina* is also an annual opportunist, found in rockpools and on unstable bottoms in less exposed conditions, whereas *S. polyschides* grows under all exposure conditions.

Other species: In Southwest Britain *Laminaria ochreleuca* is found in the shallow subtidal. This southern species has a stiff but smooth stipe and has a yellow hue particularly under water.

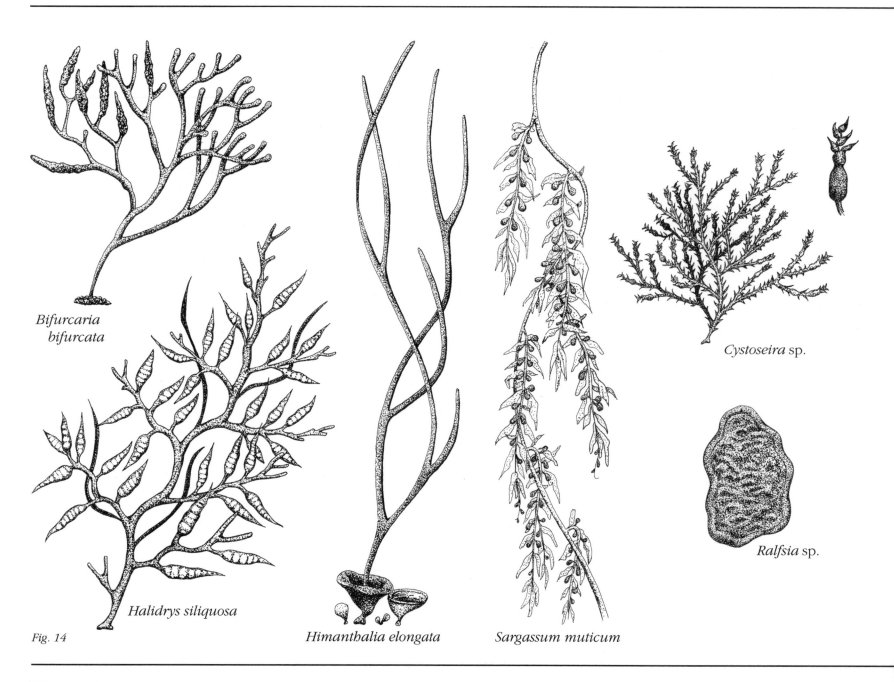

Bifurcaria
bifurcata

Halidrys siliquosa

Himanthalia elongata

Sargassum muticum

Cystoseira sp.

Ralfsia sp.

Fig. 14

These mainly large brown algae can also be locally abundant and ecologically important.

Species	Distinguishing Features	Tidal Level	Exposure	Geography
Bifurcaria bifurcata [=*B. rotunda, B. tuberculata*] III, 11	**Rounded thallus** with several branches (< 50 cm). Olive coloured, **smooth and shiny**. Usually found in *Corallina* pools.	Mid eulittoral pools	Moderately exposed to exposed	S.W., S. Devon to Pembrokeshire, extends to west coast of Ireland
Halidrys siliquosa (Sea-oak) III, 12	Flattened frond (< 1 m) with alternate branches. **Chambered air bladders at end of branchlets**. Orangy-brown.	Mid- and low-shore pools	Sheltered to moderately exposed	All coasts
Himanthalia elongata (Thong Weed) III, 13; VIII, 13	Long, thong-like thallus (< 1 m) **sprouting from "buttons"** Usually 1 or 2 main branches, sometimes more.	Eulittoral/ sublittoral fringe boundary	Moderately sheltered to exposed	All coasts
Sargassum muticum (Japanese Seaweed, Japweed) III, 14	Long, many-branched plant with small air bladders. Thin finely branched thallus, green-brown. Looks like "**washing on a line**" when main axis is held horizontally.	Sublittoral fringe and deep pools	Commonest in sheltered harbours and estuaries on boulders and stones	South coast, still spreading 1988 Cornwall to Kent
Cystoseira spp. (*tamariscifolia*) III, 15	< 70 cm. Spiky looking. Much branched irregular, yellow/brown. **Blue iridescence in water**. Small bladders form part of branches.	Lower eulittoral and sublittoral fringe pools	Moderately sheltered to exposed	South and west only, absent east coasts and rare much of Irish Sea
Ralfsia spp. VI, 15	Thin brown crusts in patches or sheets.	Lower eulittoral	Moderately sheltered to exposed	All coasts

Identification notes: *Sargassum* can easily be confused with *Halidrys*, hence its inclusion, but also with *Cystoseira* spp. (see C, p. 40; B & Y, p. 229.). It is difficult to distinguish *Cystoseira* spp. In the British Isles *C. baccata* and *C. tamariscifolia* are the most common. Hiscock (1979) is an excellent key to brown algae.

Ecological notes: *Bifurcaria* is a S.W. species which just extends into the south-west of Britain. It forms a defined zone on open rock in west Cornwall and Brittany. *Sargassum* is a Japanese "invader" which is thought to have reached W. Europe with imported Japanese oysters. It spread rapidly as it out-competed native species. *Himanthalia* is a biennial species, in the first year it is a button or toadstool shaped plant, the long thong-like reproductive frond sprouting the following year. *Ralfsia* are now thought to be the alternative phase in the life history of erect, frondose species such as *Petalonia*. *Cystoseira* can form a zone in the sublittoral fringe in the extreme south-west of Britain and on the west coast of France.

Important Red Algae – More Exposed Shores

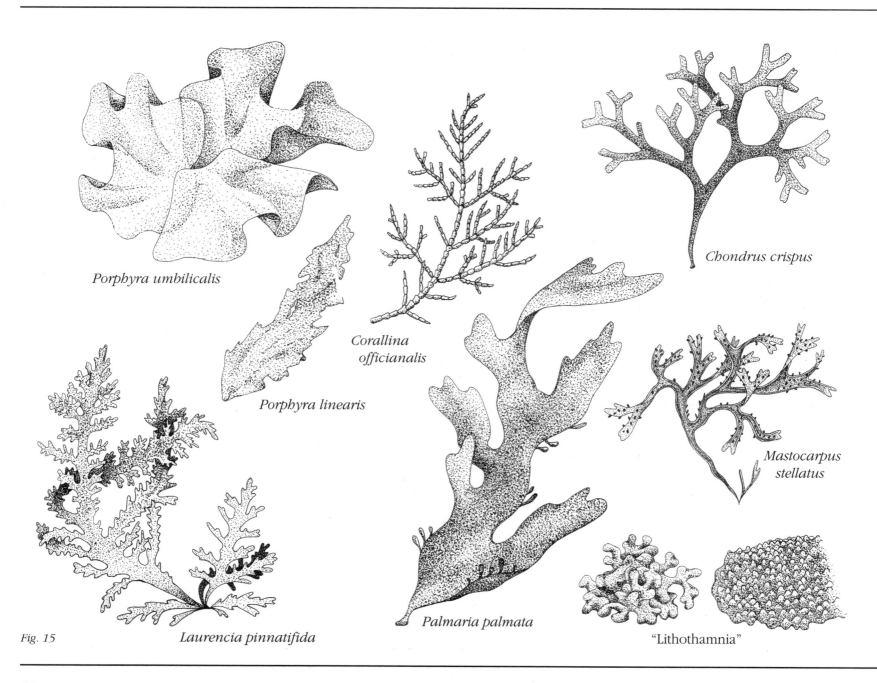

Porphyra umbilicalis

Corallina
officianalis

Porphyra linearis

Chondrus crispus

Laurencia pinnatifida

Palmaria palmata

Mastocarpus
stellatus

"Lithothamnia"

Fig. 15

Red algae (Rhodophyta) are so called because of the large amount of red-coloured photosynthetic pigments. Often, however, they can be a brown or purple colour. The species listed here are more common on exposed and moderately exposed shores.

Species	Distinguishing Features	Tidal Level	Exposure	Geography
Porphyra spp. (Laver) IV, 1 & 2	**Thallus a thin membrane** with the **texture of polythene**. No midribs or veins. Greeny-brown, brown or purple. Forms extensive sheets high on shore.	Usually littoral fringe, occasionally lower	Commonest on exposed shores	All coasts
Laurencia pinnatifida (Pepper Dulse) IV, 3	Many branched, **flattened, indented** frond. Forms patches or turfs < 20 cm tall. Alternate branches **sub-divided** into **shorter** and **shorter** branches.	Mid- and lower eulittoral	Commonest in moderately exposed conditions	All coasts
Corallina officinalis IV, 4	Purple/pink, branching plant, **stiff chalky "segments"** Forms extensive turfs about 8-10 cm high.	Mid-shore pools lower eulittoral and sublittoral fringe	Commonest on exposed shores	All coasts
Palmaria palmata [=*Rhodymenia palmata*] (Dulse) IV, 5	**Flat, oblong shaped plant** with no noticeable stalk. Sometimes sub-divided into smaller, flat, lobed offshoots from the main frond. Up to 30 cm long.	Eulittoral, sublittoral fringe, on *Laminaria* stipes in sublittoral	Sheltered to moderately exposed	All coasts
Chondrus crispus (Irish Moss, Carragheen) IV, 6	Short tufts, 10-15 cm tall. Flat frond divided dichotomously. Frond width variable. **Never inrolled or channelled.** Disc holdfast.	Lower eulittoral, sublittoral fringe	Commonest in exposed conditions	All coasts
Mastocarpus stellatus [=*Gigartina stellata*] IV, 7	As for *C. crispus* but **inrolled, channelled frond** with small blobs (fruiting bodies) on frond. Disc holdfast.	Lower eulittoral, sublittoral fringe	Commonest in exposed conditions	All coasts
"Lithothamnia" (An aggregate of several species difficult to separate in the field) IV, 8; V, 6, 7 & 13; VII 2 & 3	**A pink or mauve encrustation** covering large areas of the rock surface. Paler in pools.	Pools, lower eulittoral and sublittoral fringe	Commonest in exposure and under fucoid canopies in sheltered areas	All coasts

Identification notes: There are numerous species of red algae, some of which may be locally common. "Lithothamnia" is a collective name for several species of encrusting algae which are difficult to tell apart. Several species of *Porphyra* occur of which *P. umbilicalis*, *P. linearis* and *P. purpurea* are the commonest. When bleached, reds can appear green, or in extreme cases white. Hiscock (1986) is an excellent key to red algae.

Ecological notes: Both "Lithothamnia" and *Corallina* are impregnated with $CaCO_3$ (chalk). This is thought to give them strength and may reduce their palatability to grazers. All species, except *Palmaria* and *Porphyra* tend to be long-lived perennials. *Palmaria*, though capable of living more than one year, is an important "opportunist" species living wherever space presents itself. With the exception of the high-shore *Porphyra*, most species of red algae are found low on the shore or as understorey species below dense mid-shore canopies on sheltered shores. *Porphyra* occurs high on the shore in winter and spring, getting dried out over the summer. It will occur in "escapes" from grazing and disturbed areas on the low-shore. It has an alternate "conchocoelis" phase which burrows in shells.

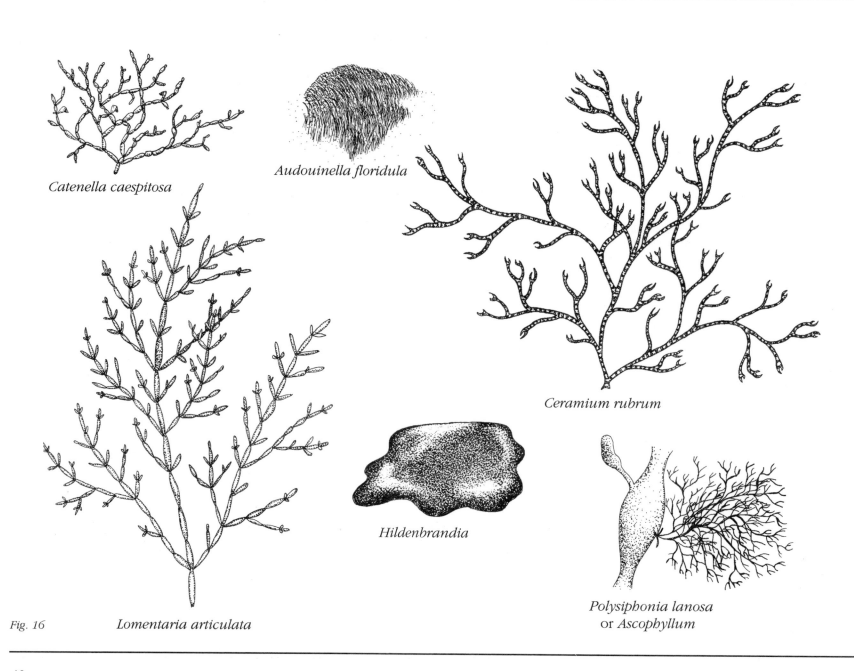

Catenella caespitosa

Audouinella floridula

Ceramium rubrum

Lomentaria articulata

Hildenbrandia

Polysiphonia lanosa
or Ascophyllum

Fig. 16

Important Red Algae – More Sheltered Shores

The species listed below are, with the exception of *Hildenbrandia* are more common on sheltered shores, particularly under fucoid canopies. A vast array of red algae occur low on the shore. Most need to be taken back to the laboratory for identification. Hiscock (1986) is excellent for this.

Species	Distinguishing Features	Tidal Level	Exposure	Geography
Catenella caespitosa [=*C. repens*] IV, 9	**Moss-like**, dark purple to black, interlaced irregularly articulated branches. < 1 cm.	Upper eulittoral, under *Pelvetia* and *Fucus spiralis* and in shade	Sheltered	All coasts
Lomentaria articulata IV, 10	Constricted, branched stem, **looks like joined beads**, purply-bright red. <15 cm.	Eulittoral, under fucoids and in shade	Sheltered to moderately exposed	All coasts
Audouinella floridula [=*Rhodochorton*] IV, 11	**Very fine red threads-fixing fine sand or mud**. Forms extensive sheets.	Eulittoral	Mainly sheltered sometimes exposed	All coasts
Ceramium spp. (*rubrum*) IV, 12 & 13	Fine banded thallus **ending in forked tips** <20 cm.	Eulittoral and sublittoral fringe	Sheltered to exposed	All coasts
Polysiphonia lanosa III, 3 & 4	Small dark red tufts **on Ascophyllum** <10 cm.	Eulittoral (not in low salinity)	Sheltered	All coasts
Hildenbrandia spp. IV, 14	Dark red, **blood-stain**-like thin crust.	Eulittoral, often in pools	Sheltered to exposed	All coasts

Identification notes: The various species of *Ceramium* are highly variable and their taxonomy is confused. It is best to lump these together, although any non-spiny plants are likely to be *Ceramium rubrum*. There are several crusts of which *Hildenbrandia* is one of the most common (see Hiscock, 1986 for key).

Ecological notes: *Audouinella* fixes sand or mud beneath *Ascophyllum* communities, but can also be found on more exposed shores at rock/sand interfaces. *Polysiphonia lanosa* is always associated with *Ascophyllum*, probably parasitically. Most of the sub-canopy species are very susceptible to high light intensities and desiccation and die off when the canopy is removed. Many of the crusts are thought to be the alternate phase in the life history of plants with an erect frondose phase (e.g. *Petrocelis* may be the crustose phase of *Mastocarpus*).

Sponges (Porifera)

Sponges are very simple animals which take in water through their porous surface and remove suspended bacteria and algae for food. Millions of flagellated cells pump the water and catch food particles. The filtered water leaves through large openings (oscula) spaced out over the surface. The skeleton of the two species named here consists of minute needles (spicules) made of silica, a form of glass fibre. The shape of the spicules is helpful in confirming identification of sponges but a microscope is needed. Other sponges have calcareous (chalky) spicules of three rays. Yet others have a fibrous skeleton of a material called spongin (e.g. the bath sponge which grows in the Mediterranean and elsewhere).

   Only two species need be considered here, both closely encrusting species coating crevices or overhangs and on more open rock under seaweed in the lower eulittoral and sub-littoral fringe.

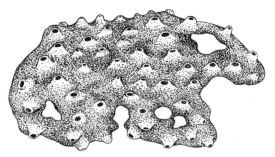

Halichondria panicea

Species	Distinguishing Features	Tidal Level	Exposure	Geography
Halichondria panicea (Breadcrumb Sponge) V, 1	Bright **green** irregular patches. 0·5 cm thick. 20-30 cm across. **Distinct** slightly raised **openings** (oscula). Can be yellow in shade. Smooth surface.	Lower eulittoral to sublittoral fringe and below	Sheltered to exposed	All coasts
Hymeniacidon perleve [=*H. sanguinea*] V, 2	**Orange,** but may be **yellow-blood red**. Up to 1 cm thick. **Indistinct** oscula . Untidy , coarse mass.	Lower eulittoral to sublittoral fringe and below	Sheltered to exposed	Not east coast, often on rocks in muddy estuaries

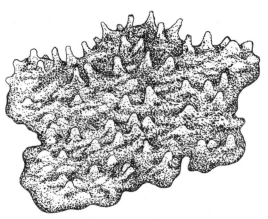

Hymeniacidon perleve

Identification notes: The shape of sponges will change with water movement. The oscula will be more pronounced and tower-like in the sub-tidal in calmer water. *Halichondria* is green due to symbiotic algae. In shade or crevices it may become yellow (see B & Y, pp. 41-41 ; C, pp. 68-75 for other species)

Ecological notes: *Halichondria* and *Hymeniacidon* are both commoner under stones and in crevices. They can form extensive sheets on overhangs and they are reasonably common on open rock beneath *Fucus serratus* and especially *Laminaria* canopies low on the shores.

Fig. 17

Hydroids are small, colonial organisms related to sea-anemones and jellyfish (Cnidaria). They consist of a number of small polyps which in some species retract into minute cups arranged either side of a stalk. There are many species, but only three are considered here as likely to be found in transects. The top two are small, 2 or 3 cm long, feather-like colonies usually attached to kelp or fucoids, the other forms pinkish, slimy tufts attached to *Ascophyllum*.

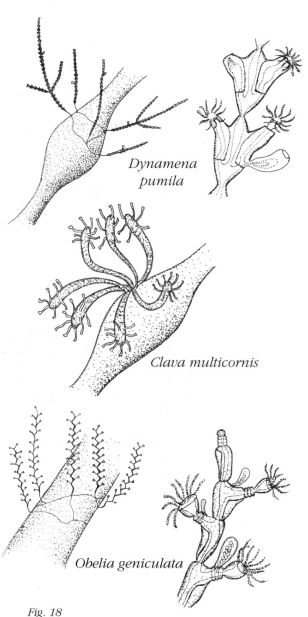

Species	Distinguishing Features	Tidal Level	Exposure	Geography
Dynamena pumila V, 3	Brown colony to about 3 cm high. Polyps in sessile cups in **opposite pairs** on the central stalk. Commonly attached to base of fucoids, esp. *Ascophyllum* and *F. serratus* where it can be dense.	Mid- to lower eulittoral	Sheltered to moderately-exposed	All coasts
Obelia geniculata V, 4	Whitish colony to 3 cm high. Polyps each on a short stalk which are **alternate** on the central stalk. Common on *Laminaria* fronds and may also occur on fucoids.	Sublittoral fringe	Sheltered to moderately-exposed	All coasts
Clava multicornis V, 5	Pink or orange, slimy, tufts 1cm long attached to *Ascophyllum*.	Mid-eulittoral	Sheltered, particularly high current areas	All coasts

Identification notes: Hayward (1988) is good for hydroids on seaweeds. B & Y pp. 45-53 and C pp.78-81 outline many common species.

Ecological notes: Hydroids usually have two phases in their life history : an attached 'polyp' stage, and a free swimming planktonic 'medusa' or jellyfish stage. They are carnivorous, catching small prey with stinging 'nematocysts' just like sea anemones.

Fig. 18

Dynamena pumila

Clava multicornis

Obelia geniculata

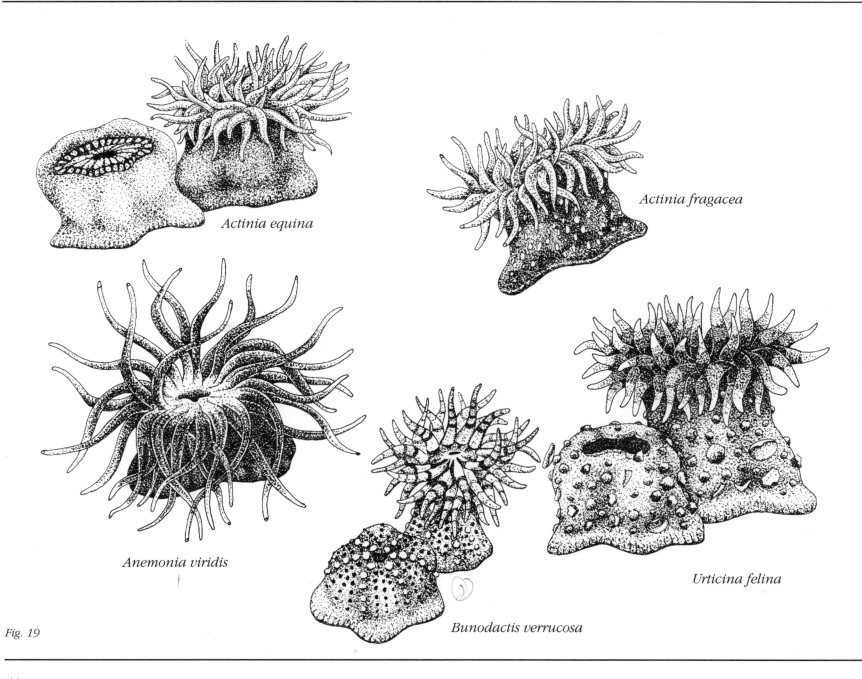

Actinia equina

Actinia fragacea

Anemonia viridis

Bunodactis verrucosa

Urticina felina

Fig. 19

Sea anemones are simple animals related to jellyfish, hydroids and corals (Cnidaria). Only 4-5 species are common intertidally, though many others occur sublittorally. One genus, *Actinia*, is reasonably tolerant to emersion and occurs very abundantly on open rocks. *Actinia* and *Bunodactis* appear as stiff jelly-like blobs when retracted but numerous short, stiff tentacles emerge and surround a central, slit-like mouth when in water.

Species	Distinguishing Features	Tidal Level	Exposure	Geography
Actinia equina (Beadlet Anemone) V, 6 & 7	Usually dark red, but green and brown varieties occur. Ring of blue tubercles below tentacles **which can be withdrawn**. 2-5 cm diameter. Looks like blob on rock.	In pools, under *Fucus*, upper eulittoral; on open rock and crevices in lower eulittoral and sub-littoral fringe	Sheltered to very exposed	Very common on all coasts
Actinia fragacea (Strawberry Beadlet Anemone) V, 8	Dark red with **green mottled** pattern. **Tentacles can be withdrawn**. Can be >5 cm diameter.	Lower eulittoral to sublittoral fringe; pools, gullies and overhangs	Sheltered to moderately exposed	South-west Britain, distribution not well known
Anemonia viridis [=*A. sulcata*] (Opelet, Snakelocks Anemone) V, 9	Can be >10 cm diameter Bright green, or dull grey-green. Often with a pinkish tinge. Purple tips to tentacles, **which cannot be withdrawn**.	Eulittoral and sublittoral fringe pools; rarely out of water	Sheltered to moderately exposed	**Absent** east coast & English Channel east of Isle of Wight
Bunodactis verrucosa V, 10	Pink with about **6 rows** of **white wartlets**. Closes tightly.	Eulittoral and sublittoral fringe, pools and crevices	Moderately sheltered to exposed	S and W coasts Ireland, Isle of Man to Channel Islands
Urticina felina [=*Tealia felina*] V, 11	Large base (10-15 cm) with squat column. Covered in sticky warts **to which sand and gravel adhere**. Bright red or green but variable colour. Tentacles totally retract.	Lower eulittoral sublittoral fringe pools and crevices	Sheltered to moderately exposed	All coasts

Identification notes: Colour morphs of *Actinia equina* have recently been raised to specific status (e.g. the *green* form has been called *Actinia prasina*, see Haylor *et al*, 1984; Sole-Cava & Thorpe, 1987).

Ecological notes: *Anemonia viridis* has symbiotic algae inside which can act as an energy source. Most anemones capture prey by using the stinging and trapping 'nematocysts' which line the tentacles and eject little darts on threads.

Other species: A variety of other species occur on the shore (B & Y p.56-61, C p.88-96), these are essentially subtidal species found in pools, crevices and under over-hangs.

Worms with hard white tubes (Serpulids)

Pomatoceros triqueter

Janua pagenstecheri

Spirorbis rupestris

Spirorbis spirorbis

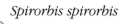

Fig. 20

Serpulids are worms which secrete and live inside a hard, white calcareous tube which is usually cemented to rock or weed. The tube is open at its widest end and, to feed, the animal extends several feathery tentacles which catch suspended particles. When retracted, the opening is plugged by an operculum of variable shape. Most of the species encountered on the shore have small, white, spiral tubes (spirorbids), but one is larger and not spiral.

Species	Distinguishing Features	Tidal Level	Exposure	Geography
Pomatoceros triqueter V, 12	Tube **not spiral**, variously snaking up to 3 cm long, 4 mm across. Tube has a notable raised ridge along its length, which extends to a point over the opening.	Lower eulittoral, sublittoral, on rocks, stones and shells	Sheltered shores	All Coasts

Spirorbids: all with a small (< 5 mm diameter) tightly spiralled tube.

Species	Distinguishing Features	Tidal Level	Exposure	Geography
Janua pagenstecheri [=Spirorbis pagenstecheri] V, 14 & 13	**Coiled anticlockwise**. Coiling usually tight and regular. Tube non-shiny. May have up to 3 ridges.	Upper eulittoral to sub-littoral and below	Sheltered to moderately exposed	All coasts
Spirorbis rupestris V, 14	**Coiled clockwise** (mouth of tube faces clockwise). **Not ridged** longitudinally, but sometimes with transverse growth lines. Usually associated with "lithothamnia". Body red.	Mid- to lower eulittoral and sublittoral fringe on open rock	Sheltered shores	West coasts, English Channel E to Swanage
Spirorbis spirorbis [=S. borealis] V, 15	**Coiled clockwise.** Tube smooth, up to 3 mm diameter, extended to **lateral flange where cemented to substrate**. Body pale greenish-brown. **On fucoids**, especially F. serratus, occasionally kelp.	Lower eulittoral and sublittoral	Sheltered shores	All coasts

Identification notes: for a full description of British spirorbids (18 species) including a useful taxonomic and ecological key see Knight-Jones & Knight-Jones (1977).

Ecological notes: On more sheltered shores, especially on "lithothamnia", under Ascophyllum and Fucus serratus canopies, Spirorbis rupestris can attain considerable cover. Barnacles are rare in these localities and spirorbids then become the main rock-dwelling (epilithic) suspension feeders. Some spirorbids have marked larval settlement preferences for particular seaweeds (Knight-Jones, 1951,1953); this enables the substrates on which they live to be used as an aid to identification.

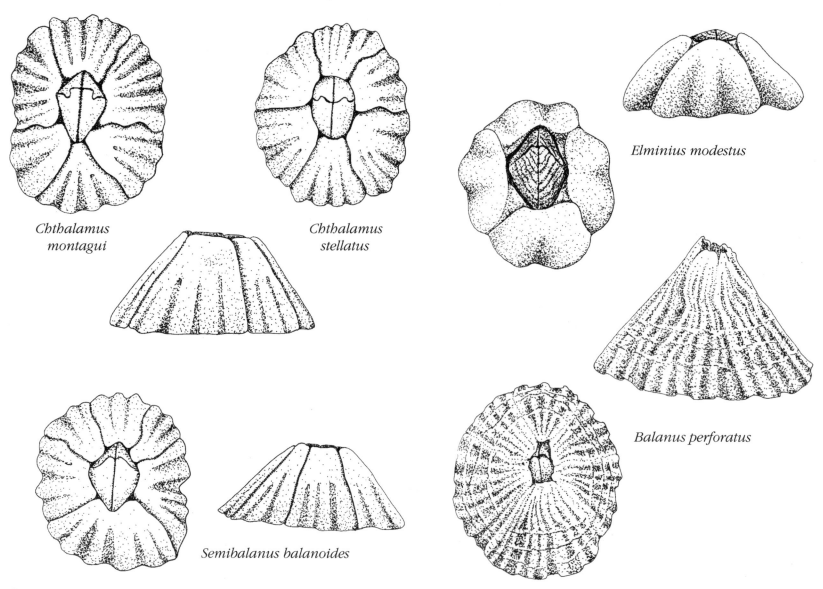

Chthalamus
montagui

Chthalamus
stellatus

Elminius modestus

Semibalanus balanoides

Balanus perforatus

Fig. 21

Barnacles are sessile crustacea, with a shrimp-like larval stage. Some confusion in identification of barnacles has been caused by the recent splitting of *Chthamalus stellatus* into *C. stellatus* and *C. montagui* (Southward, 1976). The five species listed below are the most common on British rocky shores.

Identification notes: If splashed with seawater, *C. stellatus* readily opens to reveal blue valves with a bright orange-red spot. *C. montagui* is less willing to open and when it does, it only reveals a dull browny-orange spot on its valves. The shape of the aperture is the best guide (see Southward, 1976 for key, including other species).

Ecological notes: Rainbow (1984) gives an excellent general account of barnacle biology. Filter-feeding barnacles are often the main occupiers of space on rocky shores. In the UK, their distribution patterns vary geographically. *S. balanoides*, which is dominant in northern and eastern regions, gives way to chthamalid barnacles in the South and West. *Elminius* is an Australian immigrant, first recorded in the 1940s, when it was brought here on convoys during the last war. Barnacles have a 3-4 week nauplius larval stage which feeds in the plankton followed by a non-feeding cyprid stage which shows highly selective settlement behaviour. *Semibalanus balanoides* settle in spring and early summer (March to July), the exact season varying from place to place. *Chthamalus* settle in the summer and early autumn, settlement being increasingly later and often with missing years towards their northern limit. Classic work on competition in barnacles was done at Millport (Connell, 1961a,b).

Other species: *Balanus crenatus* may be encountered very low on the shore but is essentially a subtidal species.

Species	Distinguishing Features	Tidal Level	Exposure	Geography
Chthamalus montagui (Star Barnacle) VI, 1 & 2	6 plates. **Opening angular kite-shaped.** Lines on aperture plates towards rear of barnacle and **straight** (see arrow). Browny-orange spot on valve flaps. Membranous base.	High- and mid eulittoral	Moderately sheltered to exposed	South and west coasts to Shetland, Ireland. Absent east coast south of Moray Firth to I. of Wight
Chthamalus stellatus (Star Barnacle) VI, 1 & 2	6 plates. **Opening ovate or rounded kite-shaped.** Lines on aperture near centre, and **concave** (see arrow). Bright red spot on valve flaps when open. Membranous base.	Mid- and low-eulittoral	Moderately exposed to exposed	As above but absent N Irish Sea
Semibalanus balanoides [=*Balanus balanoides*] (Acorn Barnacle) VI, 2, 3 & 4	6 plates. **Diamond-shaped aperture.** Membranous base.	High-, mid- and low eulittoral in North, mid- and low eulittoral only in S W	Varies with geography In N all shores, S W common on sheltered to mod exposed	All coasts except tip of Cornwall
Elminius modestus VI, 3 & 4	**4 plates**. Grey lines on plates on opening. Diamond-shaped opening. General pointed appearance. Membranous base.	All levels more common on mid-shore	More common in shelter and estuaries	All coasts
Balanus perforatus VI, 5	Large (1-2 cm) **purple** coloured barnacle with small aperture. **Calcareous base**. Volcano shaped.	Low eulittoral and overhangs	Moderately sheltered to exposed	S and W coasts Isle of Wight to St Davids. Absent Ireland

Bryozoans (Ectoprocta) – Sea mats

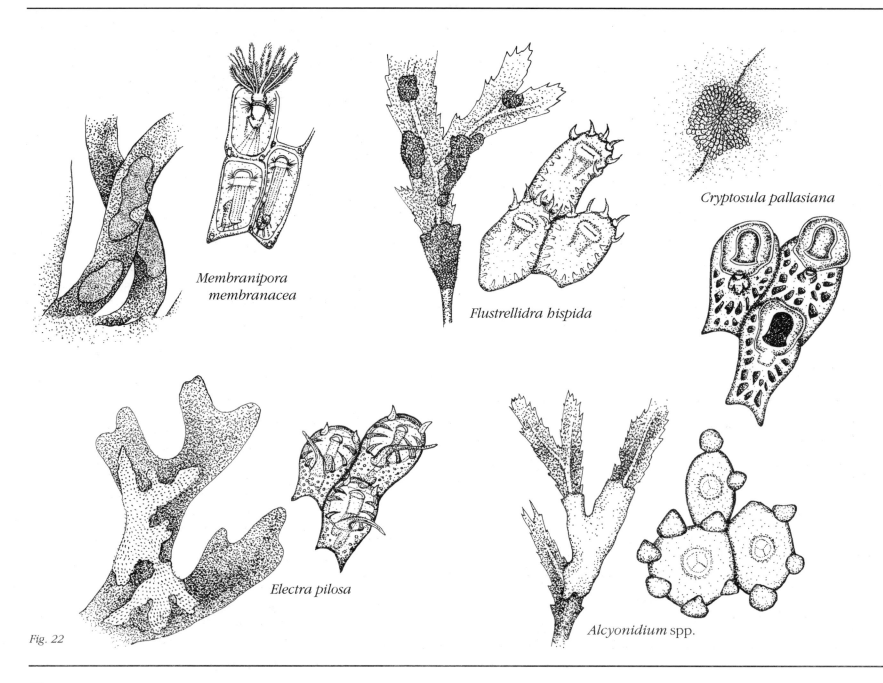

Membranipora
membranacea

Flustrellidra hispida

Cryptosula pallasiana

Electra pilosa

Alcyonidium spp.

Fig. 22

Bryozoa are mostly white, orange or brownish thin incrustations on seaweed or rock, although there are some branched or tufted forms. They consist of numerous, very small individuals (zooids) in a colony and a hand-lens is helpful to determine the shape of the zooids, which varies from species to species. Each zooid has a small retractable crown of tentacles (lophophore) with which they catch suspended food. Only 5 are considered here as common intertidally, although others may well be encountered.

Species	Distinguishing Features	Tidal Level	Exposure	Geography
Membranipora membranacea VI, 6	Forms extensive white/grey thin sheets, may be many cm across. Colony of numerous **rectangular zooids** arranged like bricks in a wall. Each zooid (< 0·5 mm) with a short blunt spine at each corner. 'Tower' zooid occasionally sticks above the colony.	Lower eulittoral and sublittoral **Very common** on *Laminaria* fronds and occasionally on other lower shore brown seaweeds	Moderately sheltered to exposed	All coasts
Electra pilosa VI, 7, 8 & 9	White irregular patches (very indented margin). Up to a few cm across. Each zooid about 0·5 mm long, **oval & surrounded** by **several short spines**, with sometimes one very long spine.	Lower eulittoral and sublittoral on red seaweeds, base of brown algae and stones, pools higher up	Moderately sheltered to exposed	All coasts
Flustrellidra hispida [Flustrella hispida] VI, 8 & 9	Brownish, lobed, **fleshy patches** a few cm across. Colony **looks hairy** ('hispida') due to horny spines on each zooid. Zooids oval, about 0·8 mm long.	Lower eulittoral and sublittoral Common on the base of *F. serratus*, *Mastocarpus* and *Chondrus*	Moderately sheltered	All coasts
Alcyonidium spp. VI, 8	**Gelatinous**, translucent patches Up to a few cm. across. Each zooid **oval or hexagonal**, about 0.5 mm long.	Lower eulittoral common on fronds of *F. serratus*, *Mastocarpus* and *Chondrus*	Moderately sheltered	All coasts
Cryptosula pallasiana VI, 10	Forms **broad, white, pink, orange patches**. Often several cm across. Each zooid oval or hexagonal with rough surface and large bell shape orifice, about 0.8 mm long.	Lower eulittoral and sublittoral, on underside of stones and shells In crevices and occasionally on *Laminaria* holdfasts	Moderately sheltered to exposed	All coasts

Identification notes: All common, encrusting intertidal bryozoans consist of small, thin, gelatinous or chalky sheets. Individual zooids, each with an orifice, can be seen using a hand lens. *Alcyonidium* spp. are difficult to distinguish in the field and are best lumped together.

Ecological notes: Most bryozoans are filter feeders needing good water movement in order to feed. They are vulnerable to desiccation and abrasion. They tend to be found on the low shore on algal fronds or in cryptic habitats (crevices, under stones) especially in areas with strong tidal currents. Several species of bryozoan can compete for space on *Fucus serratus* plants. Recently competition for food has been suggested, coupled with interactions in which faecal pellets of one species negatively affect competitors.

Other species: The four recent synopses are the definitive guides, for advanced students: Ryland & Hayward (1977), Hayward & Ryland (1979, 1985), Hayward (1985). A simpler key to bryozoans found on seaweed is provided by Hayward (1988).

Other Prominent Space-occupying Animals

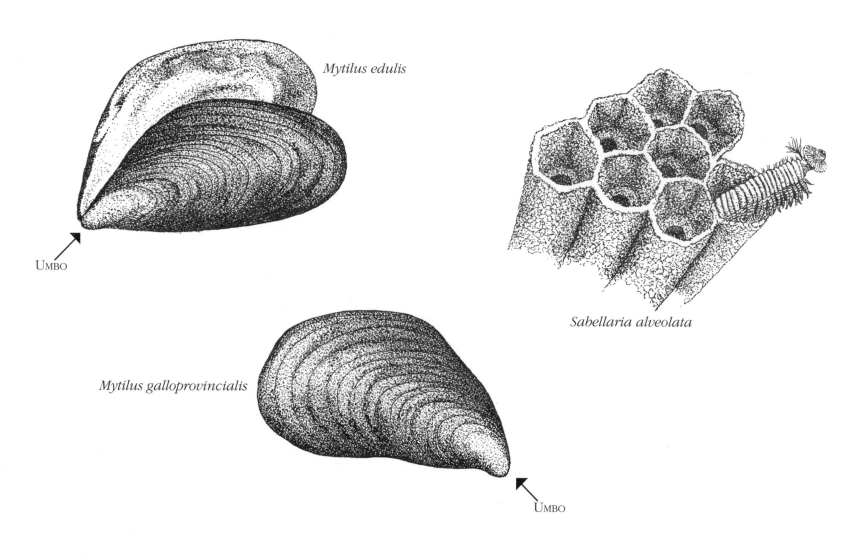

Mytilus edulis

U<small>MBO</small>

Sabellaria alveolata

Mytilus galloprovincialis

U<small>MBO</small>

Fig. 23

a) **Mussels**: There is some debate about the taxonomy of the common mussel on British shores. The two forms, *Mytilus edulis* and *Mytilus galloprovincialis*, are not easy to distinguish in the field and their specific status has been questioned (see Gosling, 1984). For most British shores it is safe to call all the mussels found *M. edulis*. Only on the south and west coasts of Britain and Ireland are *M. galloprovincialis* and *M. edulis/M. galloprovincialis* hybrids likely to be found. Both forms are illustrated, but it would be sensible to lump both species together for most ecological exercises and few would argue unduly with *Mytilus edulis* as a blanket name.

Species	Distinguishing Features	Tidal Level	Exposure	Geography
Mytilus edulis (Common Mussel) VI, 11 & 12; IV, 1	Umbo at tip, **not markedly downturned**. Top surface straight. The umbo is arrowed on figure.	Mid- and lower eulittoral	All exposures, though most common on exposed shores and in estuaries	All coasts, but patchy
Mytilus galloprovincialis VI, 12	Pointed **downturned** umbo. Top surface more rounded.	Mid- and lower eulittoral	Exposed	Ireland and S W Britain

Ecological notes: *Mytilus* are important filter-feeders and space occupiers. On horizontal surfaces they outcompete barnacles and are thought to outcompete fucoids. On exposed shores, the characteristic patches are caused by cycles of dislodgement and recolonization of mussel beds (e.g. Lewis, 1977).

b) **Honeycomb worms**: In addition to small tube worms (serpulids, see page 46), the only other segmented worm (polychaete) to be found in large quantities on rocky shores is the reef-building species *Sabellaria alveolata*.

Species	Distinguishing Features	Tidal Level	Exposure	Geography
Sabellaria alveolata (Honeycomb Worm) VI, 13	Tubes made of coarse sand grains coalesced into sheets or hummocks, often in massive **honeycomb** structures 0·5 to 1 m across.	Lower eulittoral and sublittoral fringe	Moderately exposed to exposed, or where strong currents	S and W coasts as far north as Galloway Eastern limit in English Channel at Beer. Patchy distribution

Ecological notes: *S. alveolata*'s distribution within its geographical limits is exceedingly patchy because of its special requirements. It needs high water movement from waves or currents bringing a supply of coarse sand. Larval settlement is also known to be highly patchy and spasmodic. In consequence, *Sabellaria* "reefs" tend to occur at specific sites, but do come and go, particularly as the sand/rock boundary layer they often inhabit is highly unstable. Sometimes, whole "reefs" get buried by sand movements or destroyed by storms. "Reefs" are usually between 5 and 7 years old and a predictable cycle of colonization, expansion and death occurs (Wilson, 1971; Gruet, 1986).

Identification notes: The only species *S. alveolata* could possibly be confused with is the mainly subtidal *S. spinulosa*. This sometimes occurs as groups of isolated tubes in the intertidal. On northern and eastern coasts, it sometimes forms sheets which can resemble *S. alveolata* colonies.

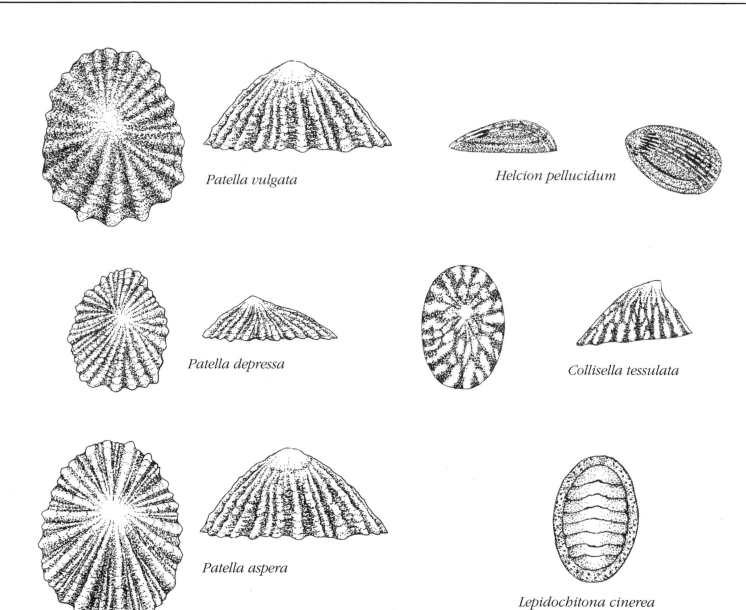

Patella vulgata

Helcion pellucidum

Patella depressa

Collisella tessulata

Patella aspera

Lepidochitona cinerea

Fig. 24

Most limpets encountered on the shores will be *Patella*. There are three British species, which can be separated quite easily after a little practice. To do this accurately, the limpets have to be removed from the rocks and examined alive. Unfortunately, they do not re-attach readily, so for conservation reasons it is often better to lump all three species together when doing a basic shore survey. It is possible, with greater experience, to tell the species apart from external shell characteristics with about a 90% success rate.

Lower down on the shore the brightly-coloured blue-rayed limpet (*Helcion*) can be found on kelps. *Collisella*, the tortoiseshell limpet is more common in the north and is usually associated with "lithothamnia".

Species	Distinguishing Features	Tidal Level	Exposure	Geography
Patella vulgata (Common Limpet) VII, 1, 2, 3, 4 & 5	**Transparent or translucent marginal tentacles**. Variable foot colour, which can be quite orangey, usually olive-green, yellow or grey. (< 60 mm)	Eulittoral most numerous mid-shore	Sheltered to exposed	All coasts
Patella depressa [=*P. intermedia*] (Black-footed Limpet) VII, 1, 3, 4 & 13	**Pronounced brilliant white marginal tentacles**. Blackish foot, chocolate rays on inside of shell. (< 40 mm)	Eulittoral, commonest around MTL	Moderately exposed to exposed	South and west coasts, Isle of Wight to Anglesey Absent in Ireland
Patella aspera [= *P. ulyssopensis*] VII, 2, 3 & 4	**Creamy marginal tentacles**, though not as obvious as *P. depressa*. **Vivid orange or pink foot**. Sometimes pinky-grey in juvenile. (< 70 mm)	Rock pools, in mid-shore if exposed, otherwise below MLWN	Moderately sheltered to exposed	South and west coasts, Isle of Wight to Humber and Ireland
Collisella tessulata [=*Acmaea tessulata*, *A. testudinalis*] (Tortoiseshell Limpet) VI, 14	Apex 1/3 length from anterior end and tilted forward. Irregular chocolate brown **tortoiseshell** markings. No pallial gills.	Sublittoral fringe and lower eulittoral pools with "lithothamnia", Often under stones	Sheltered to exposed	Northern and eastern coasts. E coast of Ireland N of Dublin, Common N of Anglesey on W. coast occasionally further south
Helcion pellucidum [=*Patina pellucida*] VI, 15	Smooth, translucent shell. Bright **lines of blue spots**. Spots not present on old limpets in holdfasts.	On *Laminaria* in sublittoral fringe and subtidal	Moderately sheltered to exposed	All coasts

Identification notes: The marginal tentacles are the only reliable characteristic. Foot colour is variable and often depends on the ripeness and sex of the underlying gonads. Both *P. aspera* and *P. depressa* have strongly ridged shells. *P. aspera* tends to be ovoid, *P.depressa* more irregular with a squarish rear edge and two pronounced rays running to the edge. See Bowman (1981) for identification of juvenile limpets. *Helcion* migrates to *Laminaria* holdfasts when older and the shell becomes thicker and loses its colour.

Ecological notes: *Patella vulgata* is the most common species. *P. aspera* and *P. depressa* are rarely found on sheltered shores. *P. depressa* rarely exceeds 50% of the population in British Isles. All are grazers and *Patella's* activities prevent seaweed growth on exposed and moderately exposed shores.

Other species: Chitons are superficially limpet-like except their shell is a series of overlapping plates; they are only locally common in the British Isles, mainly low on the shore. (see C, p. 140-141; B & Y, p.125-126). *Lepidochitona cinerea*, one of the commonest species, has been illustrated (Plate VI, 14) to give an example (see Jones & Baxter, 1987).

Topshells and Dogwhelks

Monodonta lineata

Gibbula pennanti

Calliostoma zizyphinum

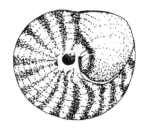

Gibbula umbilicalis

Gibbula cineraria

Nucella lapillus

Notes on dogwhelks:

Identification notes: Dogwhelks are difficult to confuse with other snails (gastropods). Similar species found lower on the shore include *Ocenebra erinacea*, young *Buccinum undatum* and *Nassarius* spp. (see B & Y, pp. 139-140; C, pp. 158-161).

Ecological notes: *Nucella* feeds by prizing open the plates of barnacles, or drilling through their shells. They also drill through mussels. On occasions, they feed on littorinids and limpets (Crothers, 1985 is an excellent summary of their biology). Recently there has been a marked decline in dogwhelks due to the influences of toxic TBT leachates from antifouling paints. These make female dogwhelks grow a penis and block the reproductive ducts, preventing breeding (see Bryan *et al.*, 1987; Gibbs *et al.*, 1987).

Fig. 25

Superficially resembling winkles, topshells tend to be covered by fine or broad lines on the shell and when the animal is withdrawn within the shell, the shell sits upright on its base, resembling a child's old-fashioned spinning top – hence their name.

Species	Distinguishing Features	Tidal Level	Exposure	Geography
Monodonta lineata (Thick Topshell) VII, 5 & 6	Underside pearly-white. **Pronounced tooth or notch on aperture**. Purple-grey colouration due to fine zig-zag lines. Top of shell often eroded. No umbilicus (hole in underside of shell). Up to 3 cm.	Mid- to upper eulittoral	Moderately sheltered to exposed	South and west, north to Anglesey and east to Portland Bill, Irish south and west coasts plus a pocket around Strangford Lough
Gibbula umbilicalis (Purple Topshell) VII, 7 & 9	**Purple-grey (or pink)** broad stripes over a grey background. **Pronounced umbilicus**. Often found on open rock. Sometimes a more greeny colour on sheltered shores. Up to 1·5 cm high.	Mid- to lower eulittoral	Sheltered to moderately exposed	South and west, north to Orkney and east to Isle of Wight All Irish coasts, except S.E. corner, French coast to Calais
Gibbula pennanti VII, 10 & 11	**Blue stripes**, very similar to *G. umbilicalis*, but **no umbilicus**. Up to 1·5 cm high.	Lower eulittoral	Sheltered to moderately exposed	French side of Channel and Channel Islands Absent Ireland
Gibbula cineraria (Grey Topshell) VII, 8 & 9; V, 7	**Thin grey lines**, more conical than *G. umbilicalis*. Umbilicus present, but not large. Rarely found in air. Up to 1·5 cm high.	Lower eulittoral and sublittoral fringe	Sheltered to moderately exposed	All coasts
Calliostoma zizyphinum (Painted Topshell) VII, 12	**Straight sided**, sharply pointed conical shell. Red streaks on a pinky-white background. All white morph is rarely found. Up to 3cm high. Body brightly coloured.	Lower eulittoral and sublittorial fringe	Sheltered to moderately exposed	All coasts
Nucella lapillus. [= *Thais lapillus*] (Common Dogwhelk) VII, 13 & 14	**Shell tapering to point**, oval opening with **grooved forwards extension**. Toothed aperture. Up to 4cm. White, or a variety of colours, particularly on exposed, *Mytilus* covered shores.	Mid- and low-shore	Sheltered to exposed shores, rarer in shelter	All coasts

Notes on topshells:

Identification notes: On the shore, shell colour, pattern and shape are helpful in distinguishing topshells from winkles. Topshells have colour bands or fine lines, are not ridged, and the shell spirals evenly as it grows. Winkles tend to be uniformly coloured (except some *L. saxatilis*), have spiral ridges (except *L. obtusata*, *L. mariae* and *M. neritoides*) and the shell spiral expands rapidly to give a larger body whorl and opening than a similar sized topshell. The operculum (horny lid that closes the shell) in topshells is circular and has many turns. That of winkles and has only 2 or 3 turns and expands more rapidly with growth.

Some confusion can arise between *Monodonta* and *G. umbilicalis*, but the presence of a tooth on the former and an umbilicus on the latter gives a clear distinction. On the French side of the Channel, *G. pennanti* and *G. umbilicalis* can be easily confused. *G. pennanti* is much bluer in colour and does not have a pronounced umbilicus.

Ecological notes: All topshells are grazers, removing bacteria, diatoms and algal sporelings from the rock surface. Their distribution changes on a seasonal basis, particularly *Monodonta*, being found lower on the shore in the colder months. *Monodonta* can be aged using lines on the shell (Williamson & Kendall, 1981). *Monodonta* and *Gibbula umbicalis* are near their northern limits and good recruitment does not occur every year.

Calliostoma keep their shells clean by regularly rubbing the foot over the shell (Jones, 1985).

Winkles

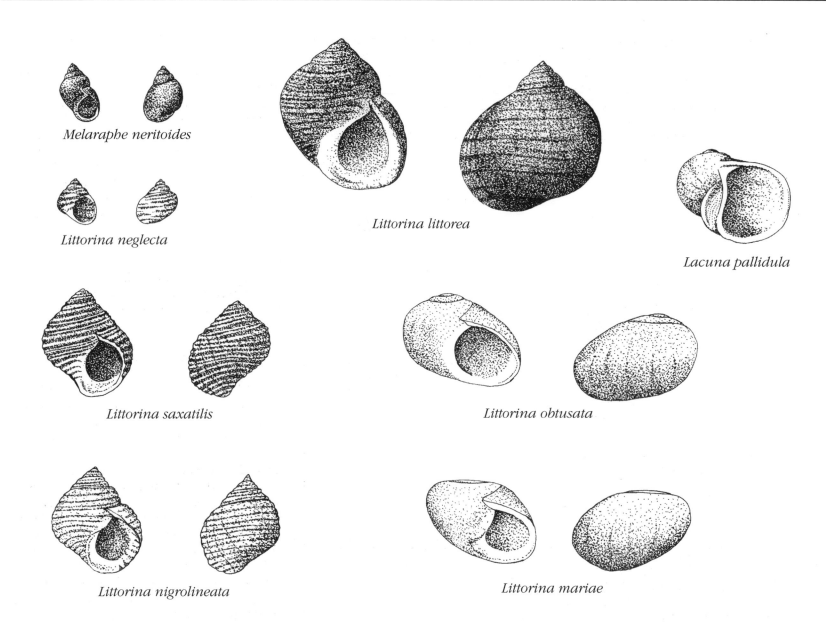

Melaraphe neritoides

Littorina neglecta

Littorina littorea

Lacuna pallidula

Littorina saxatilis

Littorina obtusata

Littorina nigrolineata

Littorina mariae

Fig. 26

The winkles are some of the commonest snails (gastropods) on the shore. There are considerable taxonomic problems with the *Littorina saxatilis* group of species and *Littorina obtusata* (sometimes referred to as *L. littoralis*) has been split into two species. We have adopted a two-stage approach: Table A splits the littorinids into four major taxa that can be readily segregated in the field, whilst Table B divides the "*saxatilis*" and "*obtusata*" groups of species into their component parts, using features which can be seen in the field using a hand lens. For most purposes, identification to the standard of the first table is adequate. One species, *L. arcana*, which can only be identified by destructive internal examination, is not segregated from *L. saxatilis*, which it closely resembles. The ability to identify the species in the second table does allow some interesting project work on niche differentiation in closely related species. No doubt opinions on the status of various littorinid species will change and the debate continue over the next few years.

Table A: basic identification

Species	Distinguishing Features	Tidal Level	Exposure	Geography
Melaraphe neritoides [=*Littorina neritoides*] (Small Periwinkle) VIII, 1	Very small (<5 mm) all black pointed shell. **Lip of opening meets body whorl tangentially.**	Upper littoral fringe	Moderate to very exposed	All coasts
Littorina saxatilis agg. [=*L. rudis*] (Rough Periwinkle) VIII, 2	Variable size (<5-20 mm) and colour. Pointed shell. **Lip of opening meets body whorl at right angles.** Lengthwise markings on tentacles.	Lower littoral fringe, upper eulittoral	Sheltered to very exposed	All coasts
Littorina littorea (Edible Winkle) VII, 6 & 15	Black (with exceedingly rare red or brown morphs). Large (< 4 cm). **Lip of opening meets body whorl tangentially.** Circular markings on tentacles.	Mid- to lower eulittoral	Sheltered to exposed	All coasts, but patchy
Littorina obtusata and *L. mariae* [=*L. littoralis*] (Flat Winkle) VIII, 3	**Blunt shell**, not having a pronounced spire (in the British Isles). Yellow, black, brown and olive morphs are the most common colours. <10-15 mm.	Mid- to lower eulittoral exposed, usually living on or amongst *Fucus* or *Ascophyllum*	Sheltered to moderately exposed	All coasts

Ecological notes: All are grazers. *L. littorea* and '*L. saxatilis*' feed on both the erect macro-algae and the micro-algal film on the rocks. *M. neritoides* feeds on micro-algae and lichens. Both *L. obtusata* and *L. mariae* live among fucoids, the former among *Ascophyllum* and *Fucus vesiculosus* and the latter primarily on *F. serratus*. These plants are both their home and their food source. *L. obtusata* actually eats the host plant whilst *L. mariae* feeds on epiphytes.

L. littorea and *M. neritoides* have widely dispersed planktonic larvae. In consequence, they tend to be genetically homogeneous, which is reflected in their uniform colour. In contrast, the *L. saxatilis* group, which are generally live-bearing (excepting the egg-laying *L. arcana*) and the egg laying *L. obtusata* and *L. mariae* are highly variable. Limited dispersal allows genetic adaptation, by selection, to localized conditions. This results in some beautiful differences in colour, shell sculpture and patterning.

Other species: *Littorina* can be most readily confused with the various *Lacuna* species (see Graham, 1971, 1988; Hayward, 1988) which are also dark and tend to live amongst seaweeds, but have an umbilicus or hole on the underside of the shell.

Table B: advanced identification

(i) *Littorina saxatilis* group				
Species	Distinguishing Features	Tidal Level	Exposure	Geography
Littorina saxatilis [=*L. rudis* + *L. arcana*] VIII, 4	Smooth or grooved. **Grooves wider than ridges**. Stripes rare, bands covering several grooves and ridges.	Lower littoral fringe, upper eulittoral	Sheltered to exposed	All coasts
Littorina nigrolineata VIII, 4	Always grooved. **Grooves narrower than ridges.** Often black stripes, restricted to the grooves.	Upper and mid-eulittoral	Moderately exposed to very exposed	All coasts
Littorina neglecta VIII, 5	Smooth or grooved. Very small (2-5 mm). **Found in empty barnacle shells**. White to mustard colour with dark brown banding.	Upper and mid-eulittoral	Moderately exposed to very exposed	All coasts

(ii) *Littorina obtusata* group				
Species	Distinguishing Features	Tidal Level	Exposure	Geography
Littorina obtusata VIII, 6	**Larger aperture** less thickening than *L. mariae*. Higher spired. < 18 mm.	Mid- and lower eulittoral (*Fucus vesiculosus* and *Ascophyllum* zones)	Sheltered to moderately exposed	All coasts
Littorina mariae VIII, 6	**Smaller heavily thickened aperture** Lower spired. < 14 mm.	Lower eulittoral (*Fucus serratus* zone)	Sheltered to moderately exposed	All coasts

Identification notes: On sheltered shores, *L. saxatilis* becomes larger with more pointed shells and sometimes it is difficult to distinguish it from *L. littorea* using shell shape. *M. neritoides* is the same shape as *L. littorea*, but there is no overlap in either the shore levels or size, so there should be no confusion. In Iceland, northern Norway and northern North America, a spired form of *L. obtusata* occurs. There is some debate about its specific status. Fortunately, it is not found in the British Isles, with the possible exception of very northern Scotland and the Northern Isles. On sheltered shores you can be reasonably sure that flat periwinkles in the *Ascophyllum* zone are *L. obtusata* and in the *F. serratus* zone are *L. mariae*. Raffaelli (1982) gives a sensible review of the ecology and taxonomy of littorinids.

NOTES:

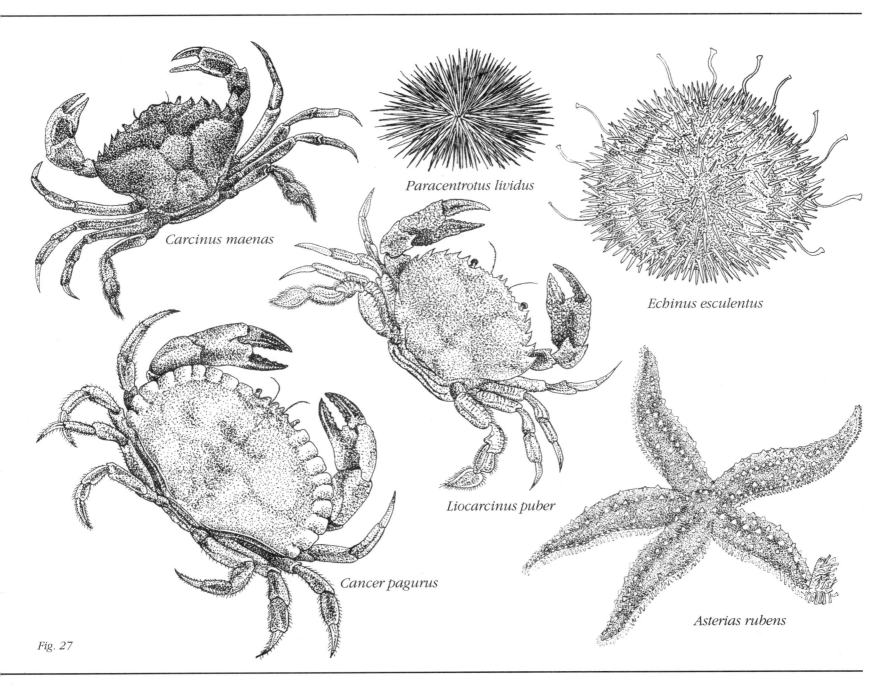

Carcinus maenas

Paracentrotus lividus

Echinus esculentus

Liocarcinus puber

Cancer pagurus

Asterias rubens

Fig. 27

Other Important Grazers and Predators

The following animals, from various phyla, are considered together as they are all important grazers (*Paracentrotus*, *Echinus*) or predators (the others).

Species	Distinguishing Features	Tidal Level	Exposure	Geography
Carcinus maenas. (Green Crab, Shore Crab) VIII, 7	0·5-12 cm. Usually dark green, can be black and have orange-red underside. Small (<1 cm) ones are often mottled white and black. **3 blunt teeth between eyes**. 5 sharp teeth on either side.	Low-shore, under stones and in pools.	Sheltered to moderately exposed, common in estuaries	All coasts
Cancer pagurus. (Edible Crab) VIII, 8	Up to 14 cm. Red-brown to purple-brown oval carapace. Large, powerful black-tipped claws. **Crinkly lobes (about 9) on sides of carapace** give "piecrust" appearance. Small ones docile when handled.	Low-shore downwards, under stones and in pools	Sheltered to moderately exposed	All coasts
Liocarcinus puber [=*Portunus macropipus*] (Velvet Swimming Crab) VIII, 9	3-12 cm. Blue joints and lines on legs, rear legs markedly paddle-like. Vivid red eyes. Velvet texture on back. **8-10 points between eyes**. Vicious, fights back when disturbed.	Low-shore to sublittoral	Sheltered to moderately exposed	All coasts, though rarer on east
Paracentrotus lividus VIII, 10	Dark green, red, brown, purply black. Smooth, solid, **sharp spines**. Often in holes or crevices. Covers test with stones, seaweeds, shells. <3cm.	Lower eulittoral rockpools into sublittoral and below	Exposed to moderately exposed	S & W coasts of Ireland, Channel Islands V. rare records in Scotland & S.W. England
Echinus esculentus (Edible Sea Urchin) VIII, 11	**Spherical, pink sea urchin** with short, blunt white spines. Obvious pedicellariae between spines. < 17cm.	Sublittoral fringe and below	Sheltered to moderately exposed	All coasts, common in intertidal in north
Asterias rubens (Common Starfish) VIII, 12	Variable colour, though **usually orange-brown**. 5 arms (sometimes 4 or 6). Can regenerate limbs. Well marked lines of spines down middle of arms' top surface.	Lower eulittoral to sublittoral and below	All exposures, most often where mussels are in the low intertidal	All coasts. Patchy distribution, commoner on shore in north & east

Identification notes: Small *Carcinus* are very common on all parts of the shore throughout the year, adults migrating offshore during winter. They can be a variety of colours, seeming to adapt to their background by the use of chromatophores. Only small, young *Cancer* will be found on the shore, the large adults living in deeper water. Although other crabs may be found under stones, the species listed are the most common (see B & Y, pp. 116-120; C, pp. 216-229 and Crothers & Crothers, 1983).

Other sea urchins which may be encountered are *Psammechinus miliaris* which tends to occur in crevices and under stones and is green in colour, and in the far north *Strongylocentrotus droebachiensis* which is primarily subtidal (see B & Y, pp. 181-183; C, pp.248-252).

The other large starfish likely to be encountered on British shores is *Marthasterias glacialis*, which is grey, covered by protruberances and can reach up to 0·5 m across. Some small brittle stars and cushion stars (e.g. *Asterina gibbosa*) may be encountered (see B & Y, pp. 178-181; C, pp. 240-247) in pools and under stones.

Ecological notes: Shore and edible crabs crack open the shells of their prey, feeding on mussels, dogwhelks and other gastropods. They also scavenge readily. Swimming crabs feed more in midwater.

Starfish feed by everting their stomachs over their prey. They feed on a variety of shore animals, including barnacles, mussels, limpets and other gastropods. Large limpets exhibit an interesting defence to starfish, raising their shells and slamming them down on the starfish's arm. If the limpet is small, it will run away rather than fight.

Sea urchins are grazers but will also eat animal encrustations. *Paracentrotus* feeds by browsing and by catching algal fragments in the spines which it passes to the mouth by tubefeet. *Echinus* is responsible for the lower limit of kelp in the subtidal in some places (see Kain, 1979; Hawkins & Hartnoll, 1983b for reviews).

NOTES

NOTES

Plate I. Zonation patterns and exposure

1 & 2. Sheltered shores.

1. General view of a sheltered shore, Menai Straits. The sublittoral fringe is not exposed on this tide. Note the virtually complete cover of seaweeds over the whole of the eulittoral. *Fucus serratus* is low down nearest the water, *Fucus vesiculosus* is a little higher, *Ascophyllum nodosum* covers the bulk of the mid-eulittoral, and *Fucus spiralis* and *Pelvetia canaliculata* are high up on the peak at upper right. The peak is topped by the yellow lichen, *Xanthoria parietina*, and just visible below the yellow is a narrow zone of the black lichen *Verrucaria maura*.

2. One of the authors, Steve Hawkins, standing against the littoral fringe and upper eulittoral of a near-vertical sheltered shore, Trearddur Bay, Anglesey. The black *Verrucaria maura* zone is near the top (head level), and below it in order are *Pelvetia canaliculata*, *Fucus spiralis* and *Ascophyllum nodosum*.

3 & 4. Moderately exposed shores.

3. View of a steeply sloping moderately exposed shore, Rhosneigr, Anglesey. The littoral fringe (black, *Verrucaria maura*) is near the top, with yellow lichens (mainly *Xanthoria*) above. The eulittoral is a patchwork of fucoids and barnacles and limpets. The fucoids are zoned from top to bottom: *Pelvetia*, *Fucus spiralis*, a few stunted *Ascophyllum*, a wide zone of *F. vesiculosus* and *F. serratus*. The sublittoral fringe is just out of the water and is dominated by *Laminaria digitata*.

4. General view of the eulittoral of a gently shelving shore showing the extensive patchiness of seaweeds, barnacles and limpets. Port St Mary, Isle of Man. (See Plate III, 10 for a view of the sublittoral fringe on a similar shore.)

5 & 6. An exposed shore, St David's Head, West Wales.

5. A general view, from nearby cliffs, at low tide on a very calm, sunny day. The wave action is due to the residual swell from the Atlantic Ocean. The eulittoral is dominated by barnacles and, remarkably, it is possible to distinguish by the colour the upper zone of paler *Chthamalus montagui* from the lower zone of darker *Semibalanus balanoides*. The littoral fringe is dominated by *Verrucaria maura* and extends several metres above the barnacles.

6. The lower eulittoral and sublittoral fringe. The rock in the sublittoral fringe is covered in pink "lithothamnia" and the kelp *Alaria esculenta*.

1   2
3   4
5   6

Plate II. Lichens, Green Algae, Wracks (part)

1-5. Lichens.

1. A rock, strictly speaking above the littoral fringe, with *Xanthoria parietina* (orange), *Ramalina siliquosa* and *Lecanora atra*.
2. *Caloplaca maritima* (centre, yellow-orange), *Xanthoria parietina* (yellow), *Lichina confinis* (bottom centre and bottom right) and *Verrucaria maura* at the upper limit of the littoral fringe.
3. Black covering of *Verrucaria maura* and blue-green bacteria at the eulittoral-littoral fringe boundary. Note the grazing activity of several *Littorina saxatilis*.
4. Extensive patches of green black *Verrucaria mucosa* in the mid-eulittoral.
5. A patch of *Lichina pygmaea* in the upper eulittoral.

6-12. Green Algae, Chlorophyta.

6. *Ulothrix flacca*, single fine, pale green threads in tufts. *Porphyra* is also present.
7. *Prasiola stipitata*, dark green.
8. *Blidingia minima*, pale green.
9. *Enteromorpha* sp., green tubular plants, here typically colonising a recently-overturned stone.
10. *Ulva lactuca* large, flat fronds, thicker to the touch than *Monostroma*.
11. *Monostroma grevillei*, thin green fronds, usually in pools.
12. *Cladophora rupestris*, here abundant in a small wet gully.

13-15 (and Plate III, 1-4). Important Brown Algae – wracks.

13. *Pelvetia canaliculata*, narrow, half-rolled fronds give the channelled appearance.
14. *Fucus spiralis*, flat, spiral frond with slight midrib. Terminal swellings are reproductive, not gas bladders.
15. *Fucus ceranoides*, narrow frond with pointed tips.

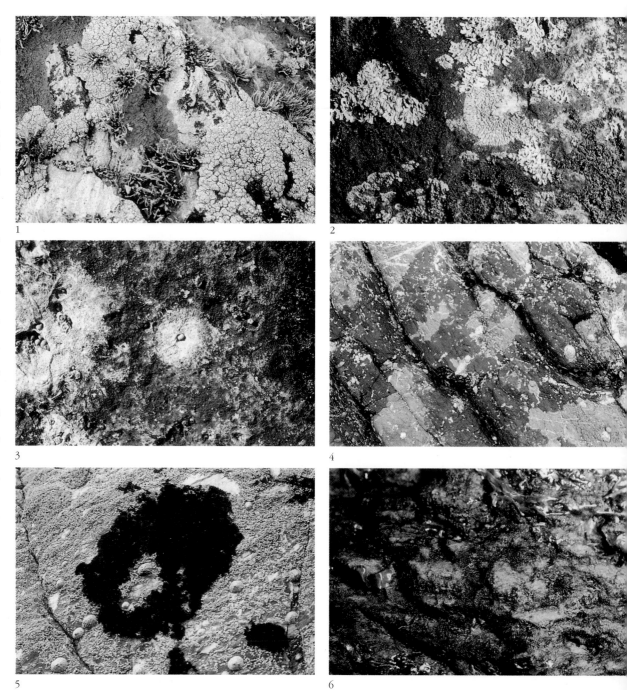

1

2

3

4

5

6

Plate III. Wracks (part), Kelps, Other Brown Algae

1-5. Important Brown Algae – wracks (continued).

1. *Fucus vesiculosus*, example from a more sheltered shore with more-or-less paired gas bladders.
2. *Fucus vesiculosus evesiculosus*, example from a moderately exposed shore showing damaged appearance and no bladders.
3. *Ascophyllum nodosum*, a prolific growth on a sheltered shore. Note the single gas bladders. The dark-green-looking tufts on *Ascophyllum* are the red seaweed, *Polysiphonia lanosa* (see also next photograph).
4. *Ascophyllum nodosum*, a stunted plant on a more exposed shore than (3). It has no bladders, and is again covered with *Polysiphonia lanosa*.
5. *Fucus serratus*, has a serrated edge, slight midrib but never swollen tips to the frond.

6-10. Important Brown Algae – kelps.

6. *Laminaria digitata* on a moderately exposed shore.
7. *Alaria esculenta* note tough midrib with fragile-looking frond either side. Centre specimen shows the reproductive "leaflets" at the base near the holdfast.
8. Several *Laminaria saccharina* plants.
9. Stranded detached specimens of *Saccorhiza polyschides* (left, with flat, wavy stipe and hollow, knobbly holdfast) and *Laminaria hyperborea* (right, three plants together, stipe is round and stiff).
10. An exceptionally low (equinoctial spring tide) low tide showing the sublittoral fringe of a moderately exposed shore. In the foreground is a zone mostly consisting of *Laminaria digitata* (collapsed). In the background, in or just out of the water, is a zone of *L. hyperborea* (stands erect due to the stiff stipe).

11-15. Other Brown Algae.

11. *Bifurcaria bifurcata*. Round thallus, regularly forking branches, in pools.
12. *Halidrys siliquosa*. In pools with pod-like, chambered air bladders at ends of branches.
13. *Himanthalia elongata*. Long, strap-like fruiting bodies grow from the centre of small mushroom-like "buttons" in their 2nd year. Several 1st year "buttons" are also visible.
14. *Sargassum muticum*, typically growing in a high eulittoral pool.
15. *Cystoseira tamariscifolia*. A stunted specimen in a pool in the lower eulittoral. The blue iridescence is evident especially on the growing tips of fronds.

1

2

3

4

5

6

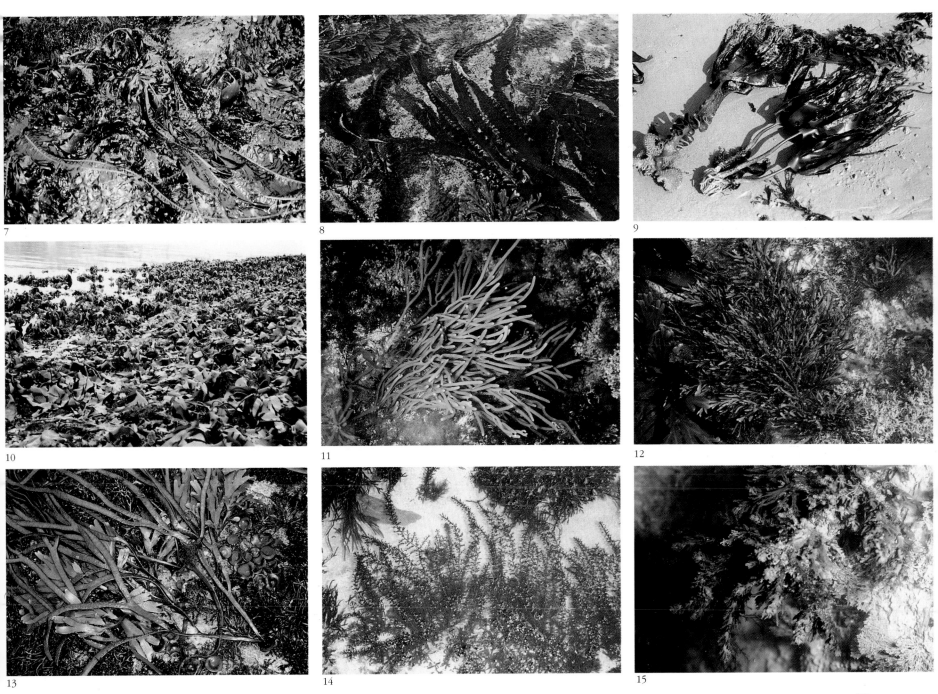

7

8

9

10

11

12

13

14

15

Plate IV. Important Red Algae

1-8. Important Red Algae – more exposed shores.

1. *Porphyra umbilicalis*, a thin purplish membrane, here growing on a clump of mussels in the upper eulittoral (compare next photograph).
2. *Porphyra purpurea* growing in the littoral fringe. It has a more greenish appearance than the species in the previous photograph due to drying.
3. *Laurencia pinnatifida* in the lower eulittoral.
4. *Corallina officinalis* growing in a pool.
5. *Palmaria palmata*.
6. *Chondrus crispus*, with *Fucus serratus* to the right.
7. *Mastocarpus stellatus*, distinguished from *Chondrus crispus* by the presence of numerous "warts" (reproductive bodies) on the surface. It can be difficult to distinguish young plants of the two species, though generally in *Mastocarpus* the frond is inrolled.
8. A healthy deep-pink growth of "lithothamnia" in the sublittoral fringe, stipes of *Laminaria digitata* are visible. Compare the paler appearance of "lithothamnia" in shallow illuminated pools, Plates V, 6, VII, 2 & 3, and the bleached "lithothamnia" visible on Plate V, 7.

9-15. Important Red Algae – more sheltered shores.

9. *Catenella caespitosa*, small moss-like plant of flattened segments, reminiscent of minute prickly-pear plant. Here growing on rock under *Pelvetia*.
10. *Lomentaria articulata*, a small plant with short, articulated segments.
11. *Audouinella floridula*, fine red threads entangled with fine sand to form a mound.
12. *Ceramium rubrum*, tuft of fine red threads with forked tips (next photograph).
13. *Ceramium rubrum*, close-up of a plant showing the curved, forked tips.
14. *Hildenbrandia* sp., a smooth dark patch on a rock.
15. *Ralfsia verrucosa*. This is a brown alga, shown here for comparison with (14).

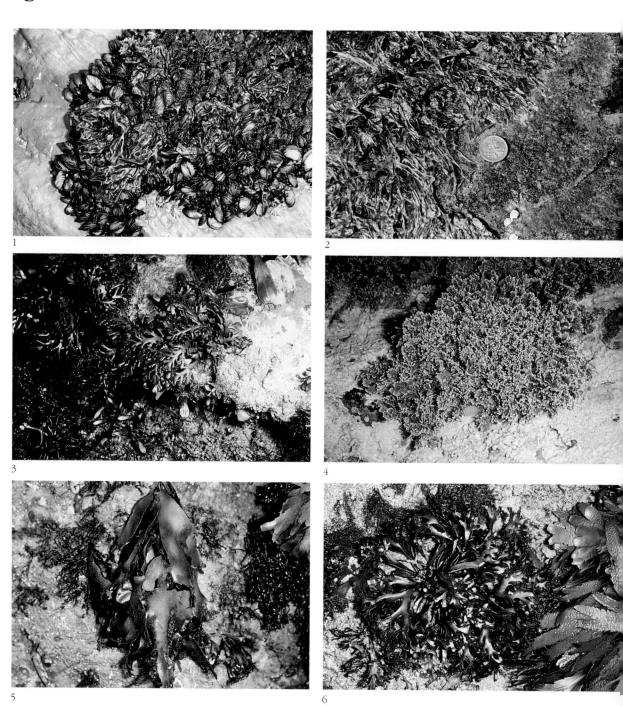

7

8

9

10

11

12

13

14

15

71

Plate V. Sponges, Hydroids, Sea Anemones, Worms with hard, white tubes (serpulids)

1-2. Sponges, Porifera.

1. *Halichondria panicea*, bright green patches with distinct raised oscula scattered over the surface.
2. *Hymeniacidon perleve*, a yellow or orange encrusting sponge with indistinct oscula.

3-5. Hydroids or sea-firs.

3. *Dynamena pumila*, short, brown stalks on seaweed with OPPOSITE, sessile, triangular cups on the stalk, here on *Ascophyllum nodosum*.
4. *Obelia geniculata*, whitish stalks with ALTERNATE, round cups each on a short stalk, here on *Ascophyllum nodosum*.
5. *Clava multicornis*, pink, jelly-like blobs (when tide is out) usually on *Ascophyllum nodosum* as shown here.

6-11. Sea anemones.

6. *Actinia equina*, an open specimen on a shallow pool encrusted with pale pink "lithothamnia" (compare Plate IV, 8).
7. Two retracted *Actinia equina*, with several *Gibbula cineraria* and white patches of bleached "lithothamnia".
8. A retracted specimen of *Actinia fragacea*.
9. Two *Anemonia viridis*, typically in a shallow pool.
10. A retracted *Bunodactis verrucosa* under an overhang.
11. A retracted *Urticina felina*, typically with attached gravel.

12-15. Worms with hard, white tubes (serpulids).

12. Several *Pomatoceros triqueter* on a large pebble.
13. Many *Janua pagenstecheri* (coiled anticlockwise) on "lithothamnia"-encrusted rock.
14. *Spirorbis rupestris* adults and newly settled juveniles (coiled clockwise), with one *Janua pagenstecheri* (coiled anticlockwise).
15. Several *Spirorbis spirorbis* (coiled clockwise) on *Fucus serratus* frond.

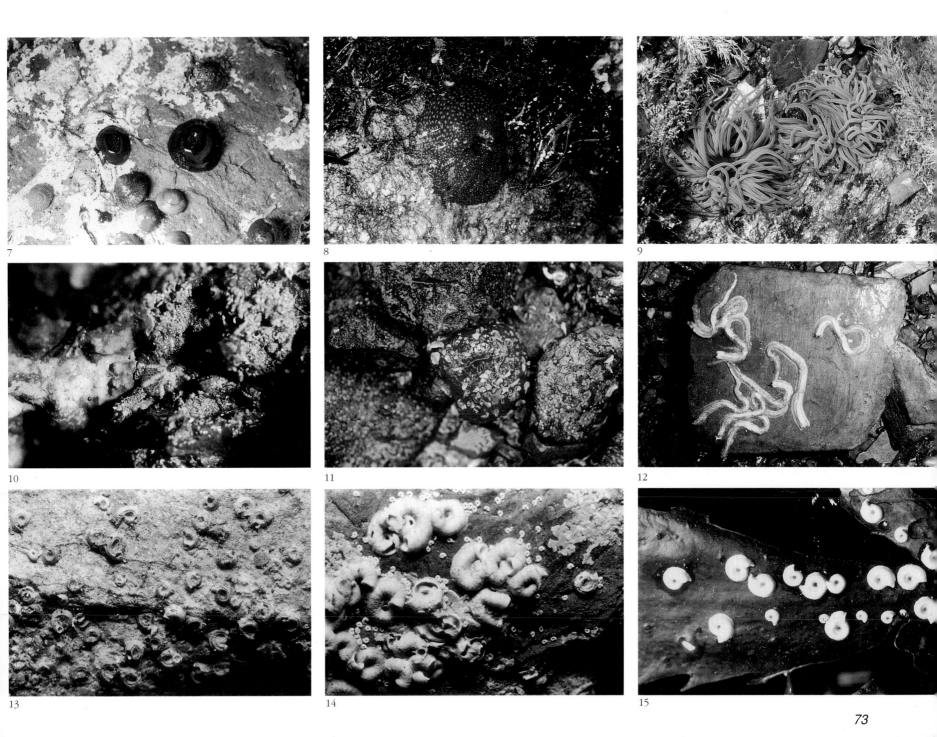

7

8

9

10

11

12

13

14

15

Plate VI. Barnacles, Sea Mats, Other prominent space-occupying organisms, Limpets

1-5. Barnacles.

1. *Chthamalus montagui*, most specimens (labelled **m**), with an angular, kite-shaped opening. A few *C. stellatus* are also present (labelled **s**), with rounded, kite-shaped opening. A *Semibalanus balanoides* is present.

2. *Chthamalus stellatus*. Rounded, kite-shaped opening, 3 specimens, with *C. montagui* (labelled **m**). Also visible are several juveniles and newly settled cyprid larvae (tiny, brown, sausage-shaped objects near the edge of the limpet, lower right) of *Semibalanus balanoides*.

3. *Semibalanus balanoides* – all the larger barnacles. Most of the group in the top right corner are *Elminius modestus* (greyer in colour).

4. A mixture of *Elminius modestus* (2 are labelled **E**; 4 plates surround the body) and *Semibalanus balanoides* (2 are labelled **S**; 6 plates, 4 larger, 2 narrow, surround the body). Both have a diamond-shaped opening.

5. Several large *Balanus perforatus*, with smaller barnacles of other species above.

6-10. Sea Mats, Bryozoa.

6. *Membranipora membranacea*, whitish, lacework patch with smooth rounded edge, here typically growing on *Laminaria digitata*.

7. *Electra pilosa*, whitish patches with very indented edge, here on *Fucus serratus*.

8. *Flustrellidra hispida* (brown, hairy), *Alcyonidium* sp., probably *A. gelatinosum* (shiny, gelatinous) and *Electra pilosa* (white, lacy), all three on a *Fucus serratus* frond.

9. Close-up views of *Flustrellidra hispida* (brown hairy) and *Electra pilosa* (white, lacy, oval openings).

10. *Cryptosula pallasiana*, hard, white or pink patch encrusting a stone.

11-13. Other prominent space-occupying organisms.

11. Mussels, *Mytilus edulis* high in the eulittoral of an exposed shore. See also next picture and Plate VII, 14.

12. *Mytilus edulis* (lower) and *Mytilus galloprovincialis* (upper).

13. A *Sabellaria alveolata* reef. (Photograph – Dr E. M. Jack).

14-15. Limpets and Chitons.

14. A tortoise-shell limpet, *Collisella tessulata* and two chitons, *Lepidochitona cinerea*, one with some white shell plates.

15. Numerous *Helcion pellucidum*, blue-rayed limpets, on the stipe of *Laminaria digitata*.

(part) and Chitons.

7

8

9

10

11

12

13

14

15

Plate VII. Patellid Limpets, Topshells, Dogwhelks, Winkles (part)

1-4. Patellid Limpets.

1. The two largest limpets are *Patella vulgata*, the medium three and very smallest one are *Patella depressa*. The second-smallest one is a young *P. vulgata*.

2. *Patella vulgata* on open rock (left), and *Patella aspera* in a "lithothamnia"-encrusted pool (right).

3. One of each patellid limpet species overturned and placed in a shallow "lithothamnia"-covered pool to show the colouration of the flesh (though this may vary a little. *Patella depressa* (left, foot dark), *P. aspera* (upper, pale foot) and *P. vulgata* (right, buff foot).

4. Close-up of the underside of the three patellid species to show the colour of the marginal tentacles. *Patella depressa* (top), *P. aspera* (right) and *P. vulgata* (bottom).

5-12. Topshells.

5. Several *Monodonta lineata*, with some *Patella*.

6. Two topshells, *Monodonta lineata* (upper) and two winkles *Littorina littorea* (lower) to illustrate the difference between winkles and topshells.

7. Four *Gibbula umbilicalis*.

8. Four *Gibbula cineraria*.

9. Two *Gibbula umbilicalis* (left) and two *G. cineraria* (right). Note the umbilicus (hole in the underside of the shell spiral) in both, but it is larger in the former.

10. Several *Gibbula pennanti* under a stone. Note the very blue colour.

11. Two *Gibbula pennanti*.

12. *Calliostoma zizyphinum*, the painted topshell.

13-14. Dogwhelks.

13. Three dogwhelks, *Nucella lapillus*, typically on a barnacle-dominated shore. A *Patella depressa* is above the *Nucella* on the right.

14. Several *Nucella lapillus* on the edge of a patch of mussels on an exposed shore. Note that shells are of several colours and patterns.

15. Winkles (also Plate VIII, 1-6).

15. Several *Littorina littorea*. Compare with *Monodonta*, Plate VII, 5 & 6.

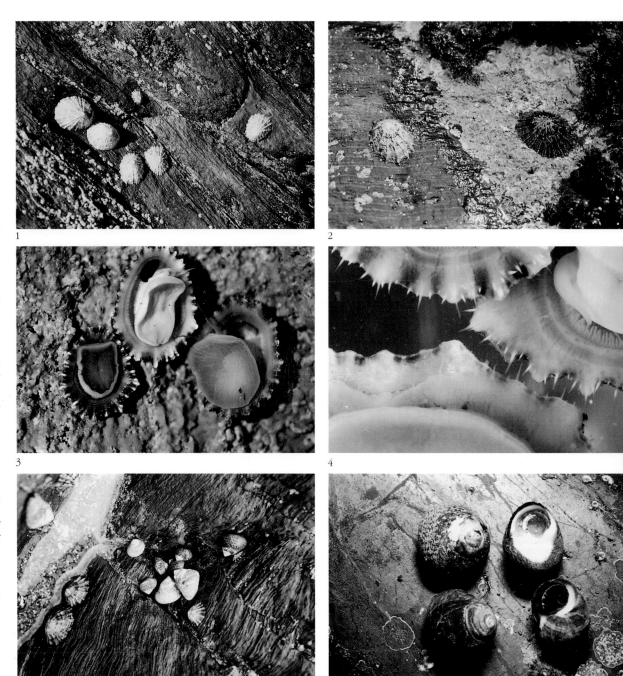

1

2

3

4

5

6

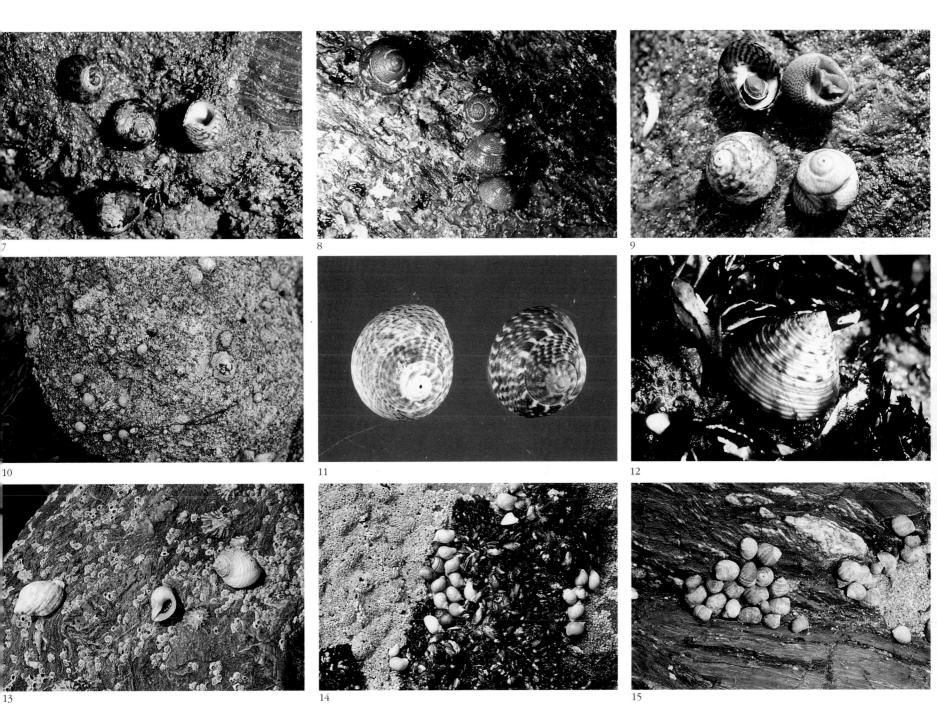

Plate VIII. Winkles (part), Other important grazers and predators, Ecology (part)

1-6. Winkles (continued).

1. Close-up view of four *Melarhaphe neritoides*, on red-coloured rock.
2. *Littorina saxatilis*, black colouration of the rock is due to the lichen, *Verrucaria maura*.
3. *Littorina obtusata*.
4. *Littorina saxatilis* (4 specimens to the left, and *L. nigrolineata* (3 specimens to the right).
5. *Littorina neglecta*, 6 specimens.
6. *Littorina obtusata* (2 specimens to the left) and *L. mariae* (2 specimens to the right).

7-12. Other important grazers and predators.

7. *Carcinus maenas*.
8. *Cancer pagurus*.
9. *Liocarcinus puber*.
10. *Paracentrotus lividus*.
11. *Echinus esculentus*.
12. *Asterias rubens*.

13-15. Ecology.

13. A rocky shore when submerged. The lower eulittoral zone of a moderately exposed shore, dominated by *Himanthalia elongata* and *Fucus serratus*. (Photograph – Dr D. Moss.)
14. An upper eulittoral shore covered in a thin film of lichens, bacteria, blue-greens and newly settled algae which make the surface dark green and very slippery. Small winkles (*Littorina saxatilis*) sheltering in the crack during low tide migrate from the crack and graze the film, showing the rock colour beneath.
15. The algal turf zone at the bottom of the eulittoral has been extended up the shore by removing limpets (and thus grazing pressure) from a 0.5 x 0.5 m square just above.

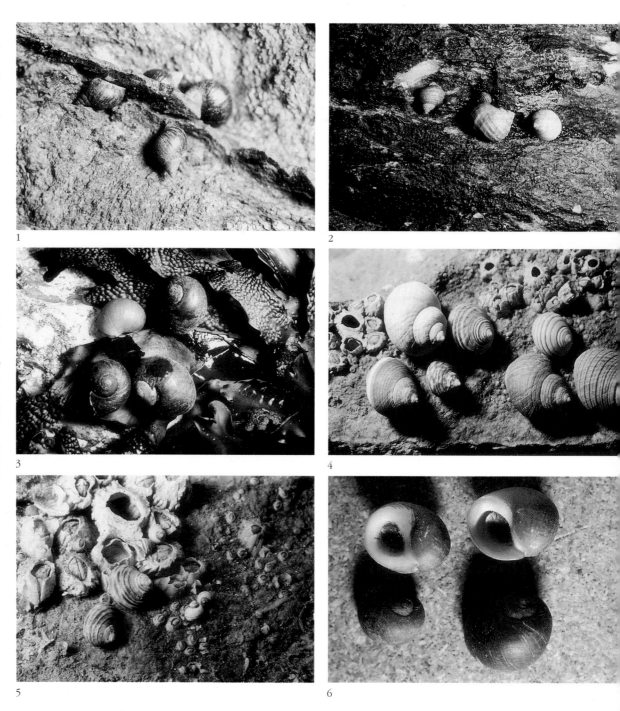

Plate IX. Ecology

1, 2 & 3. Three photos of a shore with a 2 m wide strip cleared of limpets.

1. A few weeks after removing the limpets a thin covering of ephemeral (opportunist, quick-growing) green algae has formed.

2. 9 months after – a healthy growth of *Fucus vesiculosus* has replaced the ephemeral algae in the strip.

3. 5 years after – limpets have moved into the strip and the weeds have gone, but on the shore either side the reduced grazing pressure has allowed *Fucus vesiculosus* to flourish.

4. A 1 m square on an otherwise limpet- and barnacle-dominated shore. Limpets were removed and excluded by placing a fence around to prevent subsequent inward migration of limpets. A luxurious growth of fucoids has resulted after a few months.

5 & 6. Two pictures of the same shore on the Isle of Man showing the natural change from barnacle (*Semibalanus balanoides*) domination to seaweed cover.

5. 1982.

6. 1986

Subsequently this shore has reverted to barnacle domination.

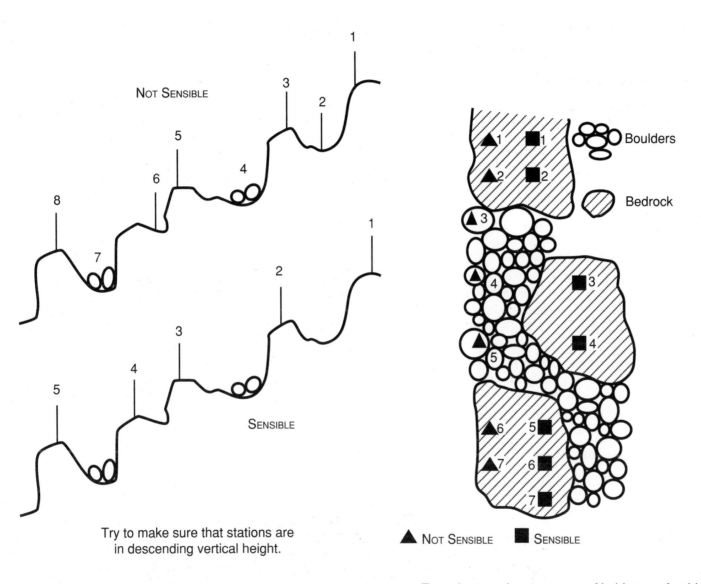

NOT SENSIBLE

SENSIBLE

Boulders

Bedrock

Try to make sure that stations are
in descending vertical height.

▲ NOT SENSIBLE ■ SENSIBLE

Try to keep to the same type of habitat, preferably bedrock. It is
legitimate to diverge slightly horizontally to do this, if the main factor
being studied is zonation.

Fig. 28

SURVEYING ROCKY SHORES

A detailed account of sampling methods for shores is given in Baker and Crothers (1986). In this section various methods are outlined and problems discussed. The methods selected depend on the aims of the study and the experience of the operators.

Reducing the Variables

The major underlying variables influencing the distribution of shore plants and animals are height on the shore in relation to tide level ("tidal height") and exposure to wave action ("exposure"). Microhabitat differences distort these major environmental gradients: pools, shade, boulders, and crevices can create low-shore conditions high on the shore; outcrops of rock can create localized havens of shelter on quite exposed headlands. Boulders can provide an unstable habitat that is constantly being recolonized by opportunist species. In most studies of distribution patterns it is best to minimize the variables by restricting the study to freely-draining, seaward-facing, solid bedrock shores. If desired, comparisons can be made between open rock and pools or crevices at the same shore level, or between pools or crevices at different tidal heights or exposures. In the latter case, similar-sized pools or crevices should be compared, as their size determines the type of environment they create. Any quick walk on the shore will show that the communities of pools are very different from adjacent open rock. This is more formally explored in the exercise on p. 94. It would be foolish to lump these two very different habitats when investigating distribution patterns.

Study of Zonation Patterns

Choosing a transect

If possible, choose a shore of unbroken bedrock which gently slopes seawards. Avoid areas near to sandy beaches which will get scoured. Boulder shores tend to be unstable and are difficult to sample, particularly if the boulders are small. They are not really suitable for introductory studies. Even on a broken shore, reasonable studies of zonation can be made provided microhabitat differences are avoided and care is taken to ensure that the various stations studied are in ascending, or descending, order of tidal height (see Fig. 28).

A **transect** is a slice down the shore taken for study. It is sub-divided into **stations**, or sampling positions, where plants and animals are counted and studied. As the main variable being explored is vertical height above low water mark, it does not matter too much if a transect is not in a strict straight line down the shore, provided the various stations are within a few metres of each other in an area of what appears to be similar wave action. Obviously unbroken transects from low to high water are best.

Levelling the shore

On unbroken slopes it is possible to survey the shore at pre-determined height intervals: splitting the littoral zone (tidal range plus an extension for splash) into 8-10 vertical stations is often appropriate. Half metre height intervals are suitable for most shores, but the intervals chosen will depend on the tidal range of the area. On broken shores with restricted areas of open rock, the shore itself determines what can or cannot be sampled. To avoid bias, take areas where the transect line intersects suitable areas of rock. On shores that are very steep or have a limited tidal range, continuous belt transects can be studied where each quadrat touches the next. For statistical reasons, however, it is preferable to have several replicates at each level.

Surveys should be made in relation to a set tidal level or **datum**. This enables comparisons between sites and occasions. The easiest way is to level from the water line at the time of low or high water. Most tide tables give the height above "chart datum" for their locality. This Admiralty reference point usually approximates to lowest astronomical tides. Provided the sea is not too rough, or the atmospheric pressure too high (lowers tides) or too low (enhances tides), this method is sufficiently accurate for most purposes. You may be fortunate and have an Ordnance Survey (O.S.) bench mark nearby. Heights on O.S. maps are given in relation to ordnance datum, i.e. mean sea level (=mean tide level). You can level down from these bench marks, but this is rarely necessary.

Levelling methods

Various methods are given in descending order of expense and accuracy.

a) *Theodolites*: You may have proper surveying equipment available. This is extremely accurate, but can be cumbersome and easily damaged by field use. Follow the instruction booklet with the level.

b) *Cowley Levels*: A split-prism level of the kind used on building sites is a cheaper and more robust alternative to theodolites. These cost in the order of £100-200 and can be bought from good builders merchants, etc. When the cross-bar attached to the levelling pole is at the same height as the window of the level, the two images in the split prisms become coincident. Again see the instruction booklet. If the money is available, these are strongly recommended.

c) *Cross-staff Levels*: These have been developed and extensively used by staff of the Field Studies Council. Fig. 29a shows their construction and operation. These are resonably accurate for most uses but only level down at fixed intervals. One tenth of the spring tide range is recommended (Baker & Crothers, 1986).

d) *Levelling Poles, String and a Spirit Level*: (see Fig. 29b): This inexpensive method is pretty accurate and particularly suitable for steep shores. The method is simple. Two graduated poles of 2-3 metres in length are placed at the two places to be levelled. A piece of string is suspended between them, and a builder's hanging spirit level (cost < £10) is used to position the string horizontally. Provided the poles are upright, the differences between the heights of the string on the poles is the vertical distance in height of the ground between the poles. A plumb line can be attached to the poles to make sure they are upright, or a small spirit level (attached or unattached) can be used.

e) *Water Bottle Method*: A bottle of water with a long tube (see Fig. 29c) can be used as a level in conjunction with a ruler.

f) *Timing Sampling*: This simple method can be used if the shape of the tide curve is known in the area (from Admiralty tide charts) or a convenient tide gauge is present in a nearby harbour. Work on an outgoing tide and mark the level of the water at set time (half hourly) intervals (stones are handy for this). The level can be worked out from the tide gauge or curve. If using a tide gauge you will have to have an observer watching this and noting the time. This method only works on calm days and a lot of waiting around is needed.

g) *Two Pole Method*: The simplest, cheapest and least accurate method is to use two cane or dowelling poles of 1·5 and 1 m (see Fig. 29d). By sighting from the 1 m pole to the top of the 1·5 m pole and hence to the sea horizon so that all three are level, gives a vertical drop of about 0·5 m. This can be used to level up or down a shore in a series of set intervals. Some red insulating tape at the top and at 0·5 m intervals is a help. You need to be able to sight to the horizon, so this method is no good in enclosed bays.

General Rules: In all cases, it is a good idea, if time permits, to determine the height of the stations in two directions, away and back to the bench mark. Very neat notes of all sightings/levellings must be made.

Assessing Abundance

Types of measures

There are two kinds of sampling: destructive and non-destructive. In destructive sampling, plants and animals are removed from quadrats (delimited by a square frame) on the shore. The organisms can be counted, or wet-weighed or dry-weighed (by putting them in an oven at between 60-100°C for 2-3 days until constant weight is

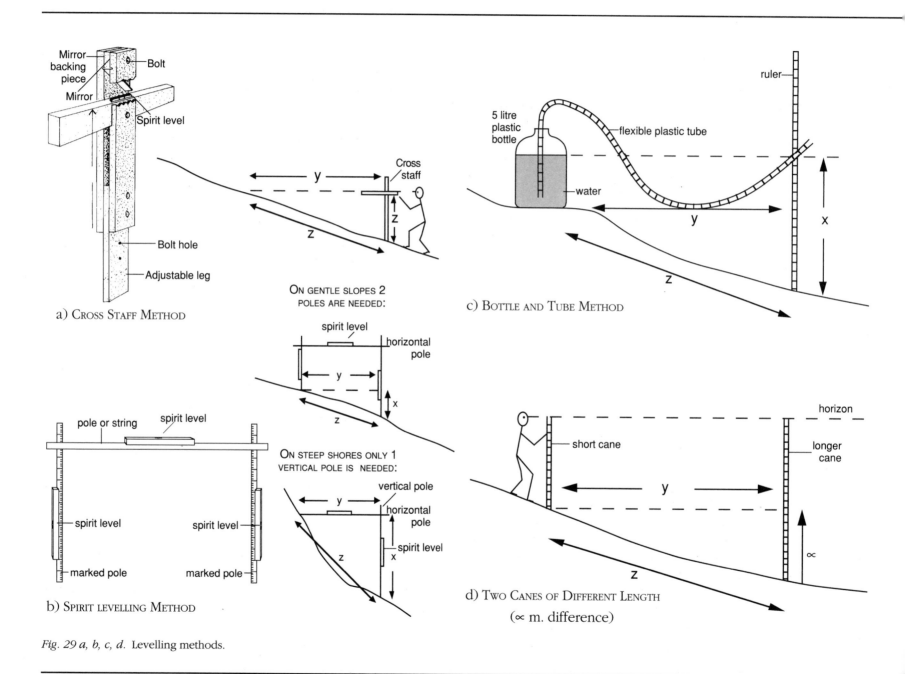

a) CROSS STAFF METHOD

ON GENTLE SLOPES 2 POLES ARE NEEDED:

b) SPIRIT LEVELLING METHOD

ON STEEP SHORES ONLY 1 VERTICAL POLE IS NEEDED:

c) BOTTLE AND TUBE METHOD

d) TWO CANES OF DIFFERENT LENGTH
(∝ m. difference)

Fig. 29 a, b, c, d. Levelling methods.

reached) to estimate biomass. This is the most accurate means of quantifying organisms and allows the sorting and identification of many of the smaller specimens found, for example, amongst barnacles, mussels and algae. However, it has some major disadvantages. Removing the organisms can severely disrupt the shore community – subsequent recolonization takes years. Not only is this disruption undesirable from a conservation viewpoint and *should be avoided whenever possible*, it also reduces the value of any subsequent study in the area sampled. Sorting of the material in the laboratory can take a long time and many specimens are damaged by the collection procedure.

Non-destructive sampling is preferable in most instances. Using this book, the major species should be identifiable in the field. Unknown species can be segregated and described as "species x", "species y"… and a sample taken to the laboratory for accurate identification.

Biomass is impossible to assess non-destructively, however, so other methods of measuring abundance must be used. Solitary animals can be counted and be expressed as **density** (no m^{-2}). This is usually done by placing a quadrat over a particular area. This is particularly suitable for mobile animals. Sessile species of animal and most plants occupy and compete for space in the intertidal. Large sedentary (slow moving) animals, such as limpets, also occupy space. Therefore **percentage cover** of available space occupied can be estimated as an ecologically meaningful measure of abundance. Shores, however, are very much 3-dimensional structures, particularly when the tide is in (Plate VIII, 13). This 3-dimensional arrangement is most developed on areas of shore with large seaweeds. These species are competing for space to trap light for photosynthesis. It is similar to the canopy of a forest. Therefore, **canopy cover** of the larger seaweeds which spread from a few holdfasts should be distinguished from **substrate cover**. Substrate cover can be defined as space occupied by ground-dwelling sessile animals (barnacles, mussels, sponges, spirorbids), smaller sub-canopy plants (the **understorey**) and encrusting plants, such as lichens and "lithothamnia". (If using % cover it is thus possible for amounts over 100% to occur!)

The distinction between canopy cover and substrate cover can become blurred: young *Fucus*, *Laminaria* and *Himanthalia* plants all occupy substrate space, often extensively. However, as they grow older, some die reducing cover of the rock surface itself. Eventually they reach a size where they start branching and forming a canopy. Other plants, such as *Chondrus*, *Mastocarpus*, and *Palmaria* can be quite large, with other species, especially encrustations, occurring beneath them. These species, however, tend to occupy space in blocks and it is difficult to segregate individual plants in any one clump. Therefore, as a general arbitrary rule, it is best to consider everything over about 15 cm as constituting canopy and anything less as occupying substrate. Encrusting plants can be treated as a separate category, but in most instances can be considered with other space occupiers.

It is possible to count non-colonial sessile animals. However, as they occupy space, for which they compete with plants, percentage cover is a more valid measure in community level studies. If, however, you are primarily interested in the distribution of different barnacle or tubeworm species, then counts may be more appropriate. Alternatively, both methods can be combined: counts can be used to find the ratio of barnacle-occupied space held by a particular species. Similarly with limpets, counts are best under most circumstances, but space occupied could be substituted or added as an additional measure. Colonial animals, such as sponges, can only be quantified non-destructively by percentage cover estimates.

Where several species of red algae, plus *Cladophora*, form a continuous sheet or turf, they can be grouped together as **perennial turf-forming algae**. Similarly, where several species of green algae (e.g. *Enteromorpha*, *Ulva*, etc.) or filamentous browns occur together, they can be lumped into the category of **ephemeral algae**. Both are sensible categories as the ecologies of the several species are similar.

SURVEYING ROCKY SHORES

Size and number of quadrats

At each level, or station, to be examined on the transect, counts can be made in quadrats. Most shores have an uneven, or patchy, distribution of species and therefore only one quadrat is likely to be inadequate, particularly if the quadrat is small.

Sampling problems are more fully explored in the class exercise on p. 98. The essential problems are what size of quadrat to use and how many repeat counts to make. The size of a quadrat depends on the size and density of the species being counted: clearly counting all the tiny upper shore winkles in a 0·5 x 0·5 m quadrat is a waste of time, just as estimating the percentage cover of 1-2 m long kelps with a 0·1 x 0·1 m quadrat is ridiculous. It is often best to use several different sized quadrats when making a quantitative community investigation. The smaller quadrats can be used to sub-sample the larger ones. Table I gives suggested quadrat sizes for various types of organisms. It is only meant as a guide! A different arrangement may be needed in your study.

The number of quadrats chosen is a compromise between time available and desire for accuracy. For general transect studies of zonation, four to five 0·5 x 0·5 m quadrats are usually sufficient, with appropriate sub-sampling for smaller organisms within each of these. On homogeneous seaweed-covered sheltered shores, or barnacle-dominated exposed shores, 2 quadrats are often enough. For most purposes, we have found a 0·5 x 0·5 m quadrat sub-divided into 0·1 x 0·1 m subdivisions to be an adequate size. A smaller quadrat can be used for counts of barnacles if desired.

Estimating percentage cover

a) *Subjective methods*: Quite good estimates of percentage cover can be made subjectively using sub-divisions of the quadrat as a guide. If working as a group, it can be worth practising to cross-calibrate. Sheets of plastic of known area (by weighing) can be used to check your accuracy. It is unwise to express the results obtained in this manner to more than the nearest 10%. Very low cover is best expressed as a "< 5 %" category. Alternatively, the

TABLE 1

SIZES OF QUADRATS RECOMMENDED FOR DIFFERENT ORGANISMS

Organism	Measure	Quadrat Size (m)
Laminaria canopies, *Ascophyllum*, *Sargassum*.	Canopy cover (percentage)	1.0 x 1.0 m (0.5 x 0.5 m will do with care).
Fucus canopies *Himanthalia*.	Canopy cover (percentage)	0.5 x 0.5 m.
Sessile animals, understorey algae, encrusting algal cover, lichens.	Substrate cover (percentage)	0.5 x 0.5 m or 0.25 x 0.25 m.
Topshells, dogwhelks, *Littorina littorea*.	No m^{-2}	1.0 x 1.0 m or 0.5 x 0.5 m (these are often very patchily distributed in crevices, pools, etc. and quadrat counts may not be appropriate).
Limpets.	No m^{-2}	1.0 x 1.0 m on sheltered shores, 0.5 x 0.5 m on mod. exposed, 0.25 x 0.25 m on exposed.
Small littorinids.	No m^{-2}	0.25 x 0.25 m or 0.1 x 0.1 m.
Barnacles, Spirorbids	No m^{-2}	0.1 x 0.1 m, 0.05 x 0.05 m or 0.02 x 0.02 m, depending on density.

cover can be split into broad categories such as 0, < 5 %, 6-10 %, 11-25 %, 26-50 %, > 50 % (see "abundance scales" in Table II).

b) *Objective methods*: i) *Frequency of occurrence.* An index of cover, but not an absolute value, can be used by recording the number of sub-divisions in which a species occurs in the quadrats and expressing this as a percentage. Clearly, the smaller and more numerous the sub-divisions are, the nearer this measure approaches an estimate of percentage cover.

ii) *Number of hits under sighting points* (see Fig. 30a). If the intersections of the string used to sub-divide the quadrat are used to "sight" on the species underneath, then the proportion of positive sightings, or hits, can be used to estimate the percentage cover. Again, the accuracy of the estimate will be increased the greater the number of intersections employed. It is best to use a thin line, such as monofilament nylon fishing line (12 lb/5 kg breaking strain is robust, but still thin enough). This method can be improved by having two layers of line to avoid parallax errors in sighting, but this is not strictly necessary. If the quadrat is on small legs, this helps sighting and allows canopy plants to be spread aside to view species underneath. This method works well for canopy cover, large clumps of understorey algae or mussels. It is not very good for thinly scattered species, such as barnacles, or spirorbids.

For looking in detail at cover of barnacles or sponges under canopies or on open rock, random dots or small holes on transparent plastic (e.g. petri dishes) or plastic sheets can be used (see Fig. 30b). This method is very good for analysing photographs. Random dots sample thinly scattered, or highly clumped, organisms better than regular arrays and there are statistical reasons for preferring them. When sessile animals are under a dense canopy with many holdfasts, subjective estimates may be the only practical method.

Twenty five sighting intersections, arranged as in Fig. 30c, are adequate for most canopy, algal turf or mussel clump estimates. The error is about 5-10 %.

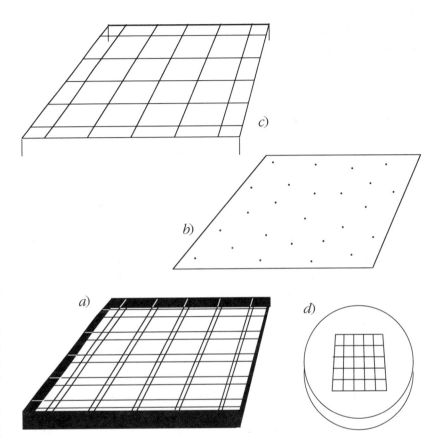

Fig. 30 a) A quadrat with a double row of cross strings for counting number of "hits" under sighting points. The double row avoids parallax errors. *b)* 25 random dots or holes on a plastic or perpex sheet. *c)* A quadrat with spikes on the corner to fit into locating holes. Note the arrangement of the strings to give 25 sighting points at the intersections. *d)* A 5 x 5 cm quadrat scratched onto a lid of a plastic petri dish – very useful for counting barnacles.

SURVEYING ROCKY SHORES

Random quadrats

At the level designated by a particular station, quadrats must be located in an unbiased manner – **not** placed over some particularly inviting clump of seaweed or cluster of snails! The best way is to use random distances either side of the transect line. Random number tables, or a pocket calculator with random number generation, can be used, or take the last three digits in the telephone directory. If many quadrats are to be sampled at a particular level, it is sometimes worth marking random distances on a piece of string or tape measure before you start. An alternative is to throw quadrats over your shoulder. This is haphazard rather than random, but done properly can be a good substitute for random numbers. Remember, in most studies of zonation you are trying to minimize microhabitat variation. Therefore, if your quadrat hits a rock pool or a large crevice reject it completely and take the next random quadrat that lands, or is placed, on open rock. This is not cheating!!

Abundance scales

An alternative to making many replicated counts in quadrats at the shore level under investigation is to use "abundance scales". These are semi-quantitative estimates of density or cover, which are assigned to 5 to 7 broad categories (see Table II, which summarizes abundance scales for various organisms). They are made in the general area of the station: usually a few metres either side and a metre or so up and down. The distance examined up and down from the station is reduced if the shore is steep, or increased if it is flat. Allocation to a particular category (e.g. "frequent" as distinct from "common") is usually helped by a few quick and approximate counts in quadrats of an appropriate size for each species (see p. 98).

Abundance scales are particularly useful for studying the abundance of a species along a stretch of coast, around an island or along a headland. The limited number of abundance categories make drawing maps easy, as different sized symbols can be used. Abundance scales, first used by Crisp and Southward (1958), were adopted by Ballantine (1961) when deriving his Biological Exposure Scale.

They have also been used extensively in pollution monitoring studies. Over the years, some workers have added categories to the 6 used originally by Crisp and Southward. The original Crisp and Southward categories are shown in bold type on Table II. We feel this is often spurious accuracy; instead of 7 or more categories, counts from replicated quadrats would be more appropriate. Additional categories also spoil the semi-logarithmic progression of the original scale.

The main advantage of using abundance scales is that they give a rapid estimate of the abundance of a particular species over a broad area, allowing an integration of abundance over an area which could only be studied by many quadrats, taking much time. This circumvents the marked patchiness of many shore organisms. It is a particularly useful approach on boulder-strewn shores or highly broken shores, where studies using replicate quadrats are often virtually impossible. The broad categories lack resolution, but the changes in abundance along vertical tidal gradients on shores are so marked that this rarely matters. The major disadvantage is that this method is prone to subjective errors. Therefore, there will be considerable inter-operator variation and hence repeatability, so essential in scientific study, is bound to suffer.

TABLE II

ABUNDANCE SCALES FOR ROCKY SHORE ORGANISMS

ALGAE
E. > 90% cover
S. 60-90% cover
A. > 30% cover
C. 5-30% cover
F. < 5% cover (zone still apparent)
O. Scattered individuals (zone indistinct)
R. Few plants – 30 min search

SMALL BARNACLES
E. >5 cm^{-2}
S. 3-5 cm^{-2}
A. > 1 cm^{-2}
(rocks well covered)
C. 0.1-1 cm^{-2}
(up to 1/3 rock covered)
F. 100-1000 m^{-2}
(individuals never >10 cm apart)
O. 1-100 m^{-2}
(few within 10 cm of each other)
R. Few found – 30 min search

LARGE BARNACLES (*B. PERFORATUS*)
E. >300 per 10 x 10 cm
S. 100-300 per 10 x 10 cm
A. 10-100 per 10 x 10 cm
C. 1-10 per 10 x 10 cm
F. 10-100 m^{-2}
O. 1-9 m^{-2}
R. Few found – 30 min search

MUSSELS AND *SABELLARIA*
E. >80% cover
S. 50-79% cover
A. >20% cover
C. Large patches
F. Scattered individuals/small patches
O. Scattered individuals, no patches
R. Few seen – 30 min search

LIMPETS
E. >200 m^{-2}
S. 100-200 m^{-2}
A. >50 m^{-2}
C. 10-50 m^{-2}
F. 1-10 m^{-2}
O. <1 m^{-2}
R. Few found – 30 min search

LICHENS, LITHOTHAMNIA CRUSTS
E. > 80% cover
S. 50-79% cover
A. >20% cover
C. 1-20% cover
(zone well defined)
F. Large scattered patches
(zone ill defined)
O. Small, widely scattered patches
R. Few patches seen – 30 min search

**DOGWHELKS, TOPSHELLS, ANEMONES
& SEA URCHINS**
E. >100 m^{-2}
S. 50-90 m^{-2}
A. >10 m^{-2}
C. 1-10 m^{-2}, very locally >10 m^{-2}
F. <1 m^{-2}, locally sometimes more
O. Always <1 m^{-2}
R. 1 or 2 found – 30 min search

LITTORINA LITTOREA
E. >200 m^{-2}
S. 100-100 m^{-2}
A. >50 m^{-2}
C. 10-50 m^{-2}
F. 1-10 m^{-2}
O. <1 m^{-2}
R. 1 or 2 – 30 min search

***MELARAPHE NERITOIDES AND
'L. SAXATILIS'***
E. >5 cm^{-2}
S. >3-5 cm^{-2}
A. >1 cm^{-2} at H.W.N.
(extending down to mid-littoral)
C. 0.1-1 cm^{-2}
(mainly in littoral fringe)
F. < 0.1 cm^{-2}
(mainly in crevices)
O. A few individuals in deep crevices
R. 1 or 2 found in 30 min search

TUBE WORMS, *POMATOCEROS*
A. >500 m^{-2}
C. 100-500 m^{-2}
F. 10-100 m^{-2}
O. 1-9 m^{-2}
R. < 1 m^{-2}

TUBEWORMS, SPIRORBIDS
A. 5 cm^{-2} on >50% of surface
C. 5 cm^{-2} on <50% of surface
F. 1-5 cm^{-2}
O. <1 cm^{-2}
R. Few found – 30 min search

Key:
E = Extremely abundant
S = Superabundant
A = Abundant
C = Common
F = Frequent
O = Occasional
R = Rare
N = Not found (all cases)

SURVEYING ROCKY SHORES

Measuring Exposure to Wave Action

General principles

Wave action is generated by wind acting over water – the longer the distance (or **fetch**) and the stronger the wind, the greater the size of the waves. Open Atlantic coasts have large widely-spaced consistent oceanic swells, often even on days without wind. This makes the open coasts of Cornwall, Devon, West Wales and Ireland ideal for surfing. In enclosed seas, such as the Irish Sea, the distance over which the wind can act is much smaller; short, steep waves which are close together result. Even in an enclosed inlet some wave action can be generated, but it will be small, steep and of short wave length. People often get seasick in harbour!

The exposure to wave action of a particular shore will also depend on the angle of the sea open to waves. Contrast the tip of an exposed headland with behind a breakwater; the headland has nearly 360° open to the sea whereas behind a breakwater may have virtually 0°. The angle open to the sea and its orientation to prevailing winds will determine the frequency or incidence of waves. Thus the angle open to the sea and the fetch over which waves have been generated determine the maximum potential exposure to wave action of a particular part of the coast. The actual wave action depends on the topography of the shore: gently sloping shores, or shores with offshore reefs dissipate wave action. Wave action is also affected by depth of water offshore: generally waves will be larger if the water is deeper.

Map-based methods (see Fig. 31)

The above first principles have led to the development of various map-based indices of exposure (details are given in Baker & Crothers, 1986). These all involve a measure of fetch (the distance of open ocean or sea) and a measure of the angle open to sea. Wright's (1981) method, outlined by Baker & Crothers (1986) also has modifying terms for minimum width of any channel to the open coast, maximum depth offshore within a mile of the coast, and the maximum

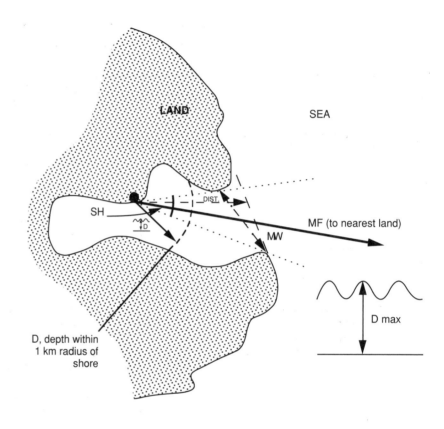

Fig. 31. An example of estimating the exposure of a shore from a map following the method of Wright (*After Baker and Crothers, 1986*).

$$MMF = MF \times (SH + 180)/MW/(Dist + MW) \times D/Dmax$$

MMF, Modified maximum fetch; **MF**, maximum fetch (distance to nearest land); **SH**, angle subtended by the sea horizon; **MW**, minimum width of channel or mouth of bay; **Dist**, distance to outer coast; **D**, maximum depth offshore within 1km of the shore; **D**max, is the maximum value in the whole survey region.

Wright's index has been slightly modified to metric units throughout (the original mixed metric and imperial). This index works well on indented coasts and allows for depth of nearshore water.

depth in the whole study area.

Additional modifications can be made to allow for strength and direction of prevailing winds. Except when working around islands or large peninsulas this is not essential. Thomas (1986) has developed more complex indices beyond the scope of this book.

A simple method is to measure the angle open to 5 or 10 km or more of sea. This index combines both a measure of 'fetch' and of 'openness'.

Direct methods

The height of waves or their upshore surge can be estimated at various points along a wave exposure gradient. Direct measurement is possible (e.g. Southward, 1953) but it is not recommended as being at best uncomfortable and usually extremely dangerous!

Wave pressures can be measured expensively using pressure transducers (Wright, 1981) or more cheaply using Denny's (1983) simple dynamometer. Wave surge (drag forces) can be measured using drogues attached to a spring balance fixed by a flexible lead to the shore (Jones & Demetropoulos, 1968; Palumbi 1984). This approach will be beyond the scope of most fieldcourses and student projects. Rates of erosion of plaster of paris have been used (Muus, 1968; Doty, 1971); in practice this is very difficult.

In addition to cost and practical difficulties, all the direct physical methods have two major scientific drawbacks. First, the single measurements made do not represent the complex of factors associated with wave exposure which affect the organisms on the shore: is average water movement, or maximum drag, or maximum length of time of calm conditions the key factor affecting your community or the species selected for study? Second, the measurements can rarely be run long enough to encompass the complete range of physical conditions encountered within the lifespan of the organisms – species like limpets may live up to 10 years.

Biological methods

Various biological exposure scales have been developed. They represent a community-level bioassay of wave exposure. The most well-known is Ballantines' (1961) developed for south-west Wales. Details of how to use this scale are available in an off-print available from the Field Studies Council. Using Fig. 4 you can quickly judge where a shore in south-west Britain lies on this scale.

The main problem with biological scales is that they strictly cannot be used to compare communities at different wave exposures. If this is done it is a circular argument: a shore is ranked as being sheltered because of its communities, but it has that community because it is sheltered. They could validly be used to compare cliff erosion! They can also be used (with reservation) to compare shores in studies of species not used as indicators in the scale. Of more doubtful validity but great convenience, is the use of the scale to compare on different shores, populations of species which are themselves used in deriving a score on the scale. This can be done as a quick rough guide, but with circumspection.

Ballantines' scale was devised for Pembrokeshire. Due to geographical changes in the balance of fucoids with barnacles and limpets with latitude (see p. 18), and the absence of some species from north and east coasts it can only be strictly applied to southwest Britain. It is possible to devise local versions (e.g. Dalby *et al.*, 1978).

Another problem which has only recently been appreciated, is that moderately exposed shores can change scale rank from 3 to 4 to 5 and back again. A good example is the switch from barnacle domination to *Fucus* cover shown on Plate IX, 5 & 6. A final consideration is that Ballantines' scale can only rank shores. The intervals between the ranks are not known. Therefore block diagrams rather than line diagrams must be used in graphically representing the results as no interpolation is possible between points (see right). Instead of 1, 2, 3, 4, 5, 6, 7, 8 the ranks could be expressed as a, b, c, d, e, f, g, h – clearly b and a half is not possible! (See Crothers, 1981, 1987)

SURVEYING ROCKY SHORES

Fig. 32. Correct (*a* – bar chart) and incorrect (*b* – line graph) way to display data when x-axis is **not** a continuous variable with known equal divisions, but simply a rank variable, like a biological shore exposure grade.

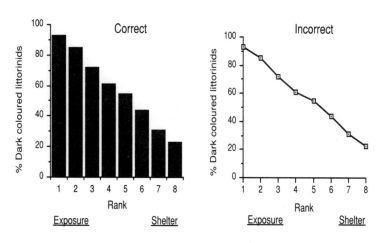

% Dark coloured *L. saxatilis* on shores at different wave exposures

All these problems aside, exposure scales are very useful in quickly classifying a shore. The biota represent a long-term integration of a complex of conditions which physical methods cannot.

Recommendations

For most purposes on fieldcourses we recommend you rank shores based on examination of a map, choosing clearly different shores. Alternatively, the simple map-based method can be used. If you rank the shore, simple non-parametric statistical methods can be used (see Hawkins, 1985 for examples).

The following class exercises have all been tried out with large groups of students (up to 70) and work well in a reasonably quick time. Obviously they should be modified to suit experience, number of students, shore type and time available. They are presented in a graduated sequence. The first exercise is to introduce the students to the shore and its biota. The second exercise is to introduce the students to a more quantitative approach to studying zonation patterns. The third exercise is to make them think about sampling problems. The fourth exercise is deliberately in outline only, as we hope the students will participate in the design of a quantitative transect exercise. The next three exercises (V – VII) can be done with both small and large groups; these focus on the **population** biology of individual species.

The exercises have been presented in the same way as a scientific paper: Introduction, Methods, Results and some questions to form the basis of a Discussion. Extras and modifications are also indicated where relevant.

CLASS EXERCISES

I. Class Exercise on Species Diversity with Shore Level and Microhabitats

Introduction

On an environmental gradient one would expect there to be less species in harsh conditions and more species in benign conditions (e.g. compare a tropical rain forest with a northern coniferous forest). A rocky shore provides a good steep environmental gradient and a range of easily identifiable plants and animals to ask the following questions:

a) How does species number vary with shore level?

b) How does species number and composition vary in different microhabitats (i.e. open rock and rock pools) at the same shore level?

This is an essentially qualitative exercise in that the presence or absence of species are noted within each shore zone and the numbers of species compared. No attempt is made to quantify any species.

Methods

a) *Equipment*: Lots of plastic bags and the means to label them.

b) *Site choice*: Select a not too steep shore with an abundance of pools at *all* shore levels. Sheltered shores are not suitable for this exercise. In many ways an exposed or moderately-exposed shore with a limited number of species is best at an introductory level. With more experienced or taxonomically competent students a rich moderately sheltered shore is fine. This exercise is best done on a spring tide.

c) *Procedure*: i) Divide the class into a least five groups. Get these working in close proximity to each other so they are essentially working on a shore of the same exposure.

ii) Divide the shore into the littoral fringe, eulittoral, and sublittoral fringe on the basis of indicator species: *Verrucaria* and winkles for the littoral fringe; the top of the barnacles or *Pelvetia* or *Fucus spiralis* in quantity (> 5% cover) for the top of the eulittoral; the top of the kelp zone for the sublittoral fringe. These zones should be demarcated on open seaward-facing rock – not pools or crevices. Further split the eulittoral into upper (*Pelvetia, Fucus spiralis, F. vesiculosus* zones, or on an exposed shore barnacles) and lower (the presence of *Fucus serratus* or a red algal turf).

iii) At each shore level search for a set time (10 or 15 minutes) and identify *in situ* (using this book) as many species of plant and animal as possible on open rock and in rock pools. **KEEP THE RESULTS SEPARATE**! Collect one or two specimens of organisms which cannot be identified in the field to identify using more specialised keys back in the laboratory. **Keep the labelled plastic bags separate**! Do not exceed your set time to enable comparisons between groups.

iv) In the laboratory identify any specimens collected. If they cannot be identified then just segregate them to 'amphipod species A'; 'isopod species B' and check with the other groups that one A is the same as another A!

Results

a) *Group results*: Each group should collate their results and produce a list of species for each microhabitat at each shore level and hence number of species of plants, animals, and the total number for each habitat, and then grand totals for each level. When counting the grand totals do not count species twice if they occur in both microhabitats.

b) *Class results*: i) Collate the class results to give the *cumulative number of species* when the group results are sequentially added together. This is best done on a large blackboard, large sheet of paper, or if available a microcomputer spread sheet. Thus a species list for group A is written down; group B tick off the species they have found and add any new species, and so on until the whole class results are amalgamated.

ii) Plot graphs of cumulative species number (y) against sampling effort (x) (minutes spent searching or numbers of groups searching).

iii) From the class species list which results from (i) above it is possible to calculate *similarity coefficients* between the various shore levels, or between microhabitats at the same shore level. This is a simple way of quantifying the similarity of community composition between two areas. One of the simplest indices to use is:

$$\frac{C}{A + B - C} \quad x \quad \frac{100}{1} \quad \text{(Jaccard's coefficient)}$$

where A = number of species in sample a
B = number of species in sample b
C = number of species in common to both

Discussion

Consider the following questions:

a) Do you have an adequate idea of the number of species in each microhabitat at each level on the shore, i) in your group results, ii) in your class results?

b) Is it reasonable to assume similar sampling effort?

c) What would happen to your results if a particular group had a specialist in amphipods (sand hoppers) working with them?

d) How does species number vary with shore level? Are there any differences in the trend between pools and open rock?

e) How useful are biological indicators in splitting the shore into major zones? Using your calculations of similarity between levels do you think a 3 zone system is justified? How could you refine the exercise to work out more objectively whether there are 3 zones?

f) Why should there be less species high on the shore?

Extras

Each species of animal can be categorised into different feeding modes (e.g. predator, grazer, suspension feeder, omnivore or scavenger) at each shore level.

CLASS EXERCISES

II. Class Exercise to Semi-quantitively Describe Zonation Patterns on Shores of Various Exposures

Introduction

This exercise is a quick way of describing semi-quantitatively the zonation patterns of the common rocky shore organisms, thereby allowing comparisons between shores of different exposures.

The specific aims of the study are:

a) What are the major patterns of zonation on the shores studied?

b) How do the abundance and zonation patterns of major organisms change with wave exposure?

c) What happens to the boundaries and positions of the three major zones (littoral fringe, eulittoral, sublittoral fringe) with increasing exposure?

To avoid the problems of patchiness explored in the next exercise the semi-quantitative abundance scales listed in Table II are used. If a large class of students is involved, some kind of cross calibration between groups in estimating abundance may be useful before the exercise is attempted. This can be done whilst waiting for the tide to go out before starting at low water.

Methods

a) *Equipment*. Each group should have: levelling poles, or cross staffs as described on p. 82; one each of 0.5 x 0.5 m and 0.1 x 0.1 m quadrats; two 1 m rulers which can be used to make a rough 1 m² quadrat by extension from the 0.5 x 0.5 m quadrat; a tape measure, or a length of string which can be measured using the metre rulers.

b) *Site choice*. Select at least three locations on a clearly defined exposure gradient, such as along a headland into a bay, or on different sides of a small island, or either side of breakwater. Try and select at least one site with a lot of seaweed cover, one with patchy seaweed and barnacles, and one predominantly barnacle or mussel covered. At each site choose a transect which has a fairly regular slope and provides an unbroken line between high and low water. Ideally shores of similar slope should be used. The transect need not be a straight line and may diverge to avoid obstacles such as large pools, outcrops of rocks, areas of sand or gullies.

c) *Species choice*. Restrict yourself to the common species in the area. Prepare a checklist or data summary sheet based on the species in this book in the following format.

SHORE SURVEY RECORD SHEET

Site: Date: Height and time of low water:

Station No:	1	2	3	4	5	6	7	8	9	10	11	12 etc
Species												
Verrucaria												
Lichina pygmaea												
etc												
etc												
etc												

d) *Procedure*: i) Levelling. Level from the sea level at the predicted time of low water (station 1) at 0.5 m vertical intervals (depending on tidal range) up the shore. Mark the stations with stones and/or cards. The use of 1 m and 1.5 m levelling poles is particularly quick and easy. Measure the horizontal distance, between each station. Thus a profile of the shore can be drawn out. By referring to tide tables the height of station 1 above chart datum can be estimated. This also allows comparison between the various transects since they all started at the same time at water level and therefore are all at the same vertical height above chart datum.

ii) *Assessing abundance*. Assess the abundance of the species of the check list in a 5 metre band either side of the transect horizontally and halfway vertically to the next station above or below. Only assess abundance for freely draining, seaward facing rock. **Ignore pools, gullies, overhangs, deep crevices etc.** Use the broad categories of the abundance scales. Quick counts with quadrats will help. In coming to an abundance category remember the aim is to get an assessment of the average abundance in the area of the transect.

• Start by estimating the cover of canopy forming algae as it lies (use the 1 m ruler to help make a 1 m² area)

• Spread the large algae aside and identify and estimate the cover on the rock of lichens, mussels, sponges, *Pomatoceros*, estimate total turf-forming algae and the main individual turf forming species, total ephemeral algae and the main individual ephemeral species.

• With the small barnacles and spirorbids, it is best to estimate their total cover irrespective of species and then identify them whilst making a couple of counts in small quadrats (5 x 5 cm, 2 x 2 cm) using a hand lens to work out the proportions of the various species. *Balanus perforatus* is large and obvious and is best considered separately.

• Mobile animals (winkles, topshells, dogwhelks, starfish) should be quickly counted and identified *in situ* wherever possible. Alternatively they can be collected into a white tray, counted and identified and then replaced in the same area. 0.5 x 0.5 m quadrats are best for large animals, 0.25 x 0.25 m or 0.1 x 0.1 m for small winkles. In the littoral fringe, *Melaraphe (Littorina) neritoides* can be counted in crevices.

• Species living on or amongst large seaweeds are difficult to quantify. Winkles or blue-rayed limpets can be counted in one or two random quadrats on undisturbed seaweeds. With other species (spirorbids, bryozoans, hydroids, green algae) note the presence or absence only at each level or use the subjective abundance scale provided.

• Always finish by having at least a 5-10 minute search in the area to see if any large but rare animals (e.g. crabs, starfish, sea urchins) are present, and to see if you can find any of the other organisms listed on the check list which may be present in small quantitites. Then if you do not find a particular species you can be fairly certain it does not occur at that level. Do not worry about species not listed on the checklist – unless they are very common in which case add them to the list.

Results

a) Plot a shore profile for each site studied.

b) The zonation patterns for selected species are best displayed as 'kite' diagrams for each site. The height above chart datum is used as the vertical scale. The abundance is expressed as units either side of this line (5 abundant, 4 common, 3 frequent, 2 occasional, 1 rare). Use the same scale to aid comparisons between sites.

c) Additionally it is a good idea to plot separately the species which indicate the various major zones (e.g. *Verrucaria*, littoral fringe; barnacles/fucoids, eulittoral; kelps, sublittoral) on the various shores next to each other. This will show if the boundaries of the zones get raised in greater exposure.

Discussion

a) Which species at which site are most sharply zoned? Why?

b) At what shore level is there maximum uplifting of zones in more exposed conditions? Why?

c) What are the major changes with increasing wave exposure? What are the factors causing these changes?

d) What are the causes of zonation?

e) How would you experimentally investigate them?

Extras

The information from these transects can be used to assess the exposure to wave action of the shore using Ballantine's (1961) biological exposure scale. If the study is done in Devon, Cornwall or West Wales (Gower to Borth) then the original scale can be used. For other areas it must be modified. A simple way of modifying the scale is to delete those species which are absent or rare in the area to be studied. It is a good exercise to apply the scale, read the original paper, (off-prints available from the Field Studies Council), and discuss the pros and cons of biological indicators versus physical measurements. Is the biological scale a circular argument?

CLASS EXERCISES

III. Class Exercise on Sampling Problems

Introduction

When using a quadrat to make counts of animals or plants, or cover estimates, two basic questions crop up: how many quadrats to use and what is the best size? Obviously, a lot depends on the aims of the study and the community or species being investigated. Most surveys involve a conflict between the desire to adequately sample an area and the time, or resources, available to do so. This is especially the case in the intertidal, where the working time is limited. The following exercise examines some of these problems on rocky shores. It addresses the following specific questions for the shore surveyed:

a) What area of the shore needs to be examined to sample most of the species on the shore? Are smaller quadrats more effective than larger ones?

b) What number of replicates are needed to give a reasonable estimate of the average density of an animal like a limpet or winkle, the average canopy cover of fucoids or the average ground cover of barnacles? What quadrat size is most suitable?

(N.B. This is also an exercise which can be done on neap tides.)

Method

a) *Equipment*: At least two 0·5 x 0·5 m quadrats, divided into four 0·25 x 0·25 m sub-quadrats should be available for each group of 4-6 students. (The quadrats can have cross-sighting strings as described on p. 86, but this is not strictly necessary as subjective estimates of cover will do for the purposes of this exercise.) If you arrive on the shore at half ebb tide the water line can be used to denote the same level throughout the study area, thereby avoiding the need to use levelling gear.

b) *Site choice*: The key to success of this exercise lies in selecting an even, gently sloping section of shore. This ensures that you are sampling essentially the same community. It is best done around mid-shore level. If a patchy *Fucus*/barnacle/limpet shore is available, this is ideal. It can also be done on barnacle dominated shores, although the species diversity will be low. It is not a good idea to do this exercise on shores with a uniform, complete cover of fucoids.

c) *Species choice*: Limpets can be readily counted and are ideal. In addition, *Fucus* canopy cover and/or barnacle substrate cover can be estimated, depending on time and shore type available.

d) *Procedure*: Organize the group to produce at least 30 large quadrats and more than 50 small ones. Get each pair of students to randomly site their large quadrats, either by throwing them over their shoulder, or using a piece of string with marks located using random number tables. In each large quadrat, record every species found, identifying them wherever possible using this book in the field. Any unknown specimens can be identified in the laboratory using a more comprehensive pocket guide or relevant keys.

Then count the number of limpets and estimate the cover of fucoids and/or barnacles. It is not necessary to identify the limpets, fucoids and barnacles to species when quantifying them, but it is worth identifying them for the species list for each quadrat. While the quadrat is still in place, carry out exactly the same procedure for the top-left sub-quadrat. Repeat as necessary to yield the desired number of quadrats – you will have to do several extra small sub-quadrats. Remember to re-throw the quadrat before doing each sub-quadrat as the four adjacent sub-quadrats are not randomly distributed with respect to each other. **It is essential that the records for each large and small quadrat are kept separately!** Also, any specimens for further identification should be kept in separate labelled bags so that their contents can be ascribed to a particular quadrat.

e) *Estimated time*: Depending on group size < 1 hr in field, 1 hr laboratory identification, 1 hr result collation and work-up.

Results

a) *Species number versus area:* For the class results prepare a table in the following form. Enter new species on the list and tick off species already found.

TABLE A Cumulative number of species found with increasing numbers of 0.5 x 0.5 m quadrats

Species	Quadrat No.				
	1	2	3	4	…etc.
Fucus vesiculosus	√	√		√	
Patella vulgata	√		√		
Semibalanus balanoides	√	√	√	√	
Laurencia spp.		√		√	
Littorina littorea			√		
Mytilus edulis				√	
etc…					
No. of species	3	3	3	4	etc.
Cumulative No. of species	3	4	5	6	etc.

You can then plot the cumulative number of species against increasing sample area (= number of quadrats).

b) *Estimates of density or cover.* For the class results prepare a table in the following form for each size of quadrat and each species.

TABLE B Sequential estimates of mean number of limpets and standard deviations with increasing sample size.

Quadrat size : 0.5 x 0.5 m	Quadrat No.						etc.
	1	2	3	4	5	6	…
Number	4	5	3	2	6	6	…
Estimated mean/quadrat =	4.0	4.5	4.0	3.5	4.0	4.3	…
Estimated Standard Deviation (S.D.) =	…	0.71	1.00	1.29	1.58	1.63	…

The sequential estimates of the mean per quadrat can then be plotted against increasing number of samples. The standard deviation(s) can be plotted as an error bar either side of the mean. It is probably best to estimate the mean and standard deviation for each new quadrat up to 10 or 15 and then every five quadrats subsequently. Adequate sampling can be simply judged by inspection, noting when the estimates stop fluctuating.

Statistical considerations

a) Species number versus area: a logarithmic transformation of the numbers of sampling units often results in a straight line relationship.

b) Comparing estimates of density with different sized quadrats can present problems. The best approach is to plot each density separately, as mean number per quadrat ± standard deviation, but adjust the scale of the graphs so that they represent number m^{-2}. This problem does not occur with % cover estimates.

c) Further statistical analyses can be done with these data. The

CLASS EXERCISES

variance/mean ratio (variance=square of standard deviation, s^2) can be used to test whether the distribution patterns are regular, random or clumped. A variance/mean ratio of 1= a random distribution, >>1 a clumped or contagious distribution, a perfectly regular distribution = 0 and regularly distributed species are often <1. The pattern found depends on the quadrat size used. For further details see Elliot (1977). Alternatively frequency histograms of the number of quadrats with 0, 1, 2, 3 etc. limpets can be plotted and tested against a poisson distribution to detect departure from random distribution.

Discussion

Consider the following questions:

a) Does the number of species level out after a certain sample area? Would you expect to find all the species or would you continue to find the odd rare one?

b) Why do you get the odd spurt in increase in species number producing a step-like graph?

c) What would happen if you re-randomized the order of quadrats and plotted the graph of species number against sampling area again?

d) What is most efficient for sampling all the species: a few large quadrats or many small quadrats? Compare your two graphs.

e) Does your estimate of limpet density, barnacle cover or *Fucus* cover flatten out? What does the residual level of fluctuation represent? What seems a good number of samples to estimate the abundance of each?

f) If you have plotted a frequency histogram of number of limpets per quadrat, what shape is it? Would you expect a normal distribution? What shape would it be if the limpets were found in a much more clumped manner on the shore?

Extras

It is also possible to plot scattergrams of limpet numbers against fucoid cover or barnacle cover for each individual quadrat to see if there are any correlations in distribution. If desired, parametric correlation coefficients (Pearson's) can be calculated. As much of the data from counts in quadrats are unlikely to be normally distributed, a non-parametric method, such as Kendall's or Spearman's rank correlation coefficient is probably more appropriate (see any standard statistics book for details).

NOTES

IV. Quantitative Exercise on Zonation

Introduction
This study has the same aims as exercise II but uses more rigorous methods.

Methods
As in exercise II designate 8-12 stations throughout the tidal range. At each station sample 4-5 randomly placed 0.5 x 0.5 m quadrats. First estimate the abundance of canopy forming species, then push these aside to estimate cover of understorey algae, encrusting forms and cover of sessile animals such as barnacles. If desired, one or two subsamples can be made in each large quadrat using a small quadrat or a subdivision of a quadrat (0.05 x 0.05 m or 0.1 x 0.1 m) to count numbers of barnacles or spirorbids. Similar subsampling can be used for numerous small littorinids. Count any large animals in the 0.5 x 0.5 m quadrats. If you wish, a short general search (10 minutes) can be made at each station to count the abundance of large predators such as crabs or starfish. Their abundance can be expressed as numbers caught per unit time. If any of your random quadrats hit a pool or a deep crevice, reject it and locate another on the open rock.

Results
Express your results as mean cover or average numbers per metre squared (four quadrats make this easy). The standard deviation can be calculated to give a measure of variation. Strictly speaking for percentage cover data, an arcsine transformation (angular transformation) needs to be applied as percentages are rarely normally distributed. For most teaching purposes this is not necessary. Use your results to construct kite diagrams to show the zonation patterns as in exercise II.

Discussion
a) If you have performed exercise II on the same shore, how much extra information do you get using this more refined approach?
b) Are you happy with the number of replicates? (See exercise III.)

Extras
The population size structure of common species such as barnacles, limpets, dogwhelks, topshells can also be measured as part of this exercise. Comparisons between pools and open rocks can be made at the same shore level.

CLASS EXERCISES

V. Distribution, Population Structure and Shell Shape of Limpets

Introduction

The limpet, *Patella vulgata* L., is common on shores of all exposures in the eulittoral zone from high to low level. Limpets have planktonic larvae and maximum spat (juvenile) settlement is reported as occurring just above MLWN level on the shore or in pools and other moist places such as under seaweed. Thus highest numbers of small limpets might be expected to be found in the lower eulittoral or near pools or weed.

Though conical, the shell of limpets may vary a little in shape. Shells of limpets higher up the shore are reported as being taller than those from lower down the shore. The tallness might also vary with the degree of wave exposure of the shore.

This project investigates variation in abundance, population structure (numbers of different sized limpets) and shell shape in relation to shore level. In the South and West it can be extended to include *P. depressa* and *P. aspera*.

Methods

Select suitable transects on one or more shores. If there are enough students, shores of different exposure can be compared. Choose shores of obviously increasing exposure (e.g. a series from inside a bay along a headland). Alternatively assess the exposure value (1-8) using Ballantine's (1961) scale (see p. 91). At regular vertical intervals (4 or 5 will suffice) levelled relative to low water, count and measure the shell length and shell height of each limpet in a 0.5 m x 0.5 m quadrat. Measure to the nearest mm. Be careful to measure all limpets including the small ones. Repeat for about 5 quadrats at each level or until about 100 limpets have been measured at each level. Note also the relative abundance of seaweeds and proximity to pools.

NOTE ON CONSERVATION

The measuring can be done without removing the limpets from the rock. In attempting to distinguish between the three species of *Patella*, it may be necessary to remove each limpet. If this is done, please wet each one in sea-water from a nearby pool and place it back on its home. Hold it there for a few seconds until it has re-attached. Avoid removing limpets whenever possible.

Results

a) Express the numbers of limpets against shore level using kite diagrams. Where are there most limpets?

b) Plot length-frequency histograms for each level. Use 5 mm size classes (<5 mm; 6-10 mm; 11-15 mm; 16-20 mm and so on). Try different size classes (say 3 mm and 10 mm). Which size class is 'best'? Where are there more small limpets?

You could plot the data using 1 mm size classes and see if there are any peaks in the data. These peaks could be year classes, limpets of the same age being of a similar size.

c) Calculate the ratio of average limpet height against average limpet length (H/L) for each shore level. Alternatively plot on a graph length (x axis) against height (y axis) for each individual limpet and fit a line by eye through the data. Better still use regression analysis (using a calculator with 2-way statistics) to find the intercept and slope and hence the equation of the line. The slope is the same as the ratio of height to length. Tabulate the ratio (slope) against shore level. Where are the limpets most conical? Standard statistics books will tell you how to compare regressions. With two samples a form of t-test can be used.

Discussion

a) Do your results support the suggestion that limpets settle in the lower eulittoral or near pools or weed?

b) What advantages would a conical shell confer to a high shore limpet? (Clue: how would the ratio of shell perimeter to biomass differ?) What advantage would a flatter shell give to a low shore limpet? A suggested (but not proven) mechanism of producing a taller shell is that in higher shore limpets the mantle which secretes the shell is pulled inwards for longer periods due to tidal emersion. Can you think of any alternative explanation?

c) Do limpets change shape as they get older?

d) If you have compared two or more shores, can your results be related to differing shore exposure?

e) How could you experimentally test some of these ideas?

(Refs: Ebling *et al.*, 1962; Moore, 1934; Fretter & Graham, 1962; Bowman & Lewis, 1975)

CLASS EXERCISES

VI. Shell Form of the Dogwhelk, *Nucella lapillus*

Introduction

Dogwhelks lay fixed, vase-shaped, egg capsules attached to stones and the young hatch as miniature adults. The young thus stay in the same area as the parents and a population may have become genetically isolated if no new individuals come from another area. Over many generations this limited gene flow coupled with selection leads to considerable differences between shores. Variation in shell colour and pattern may occur within populations. Shell form may also vary between populations, which is what this project investigates. (In contrast, the winkle *Littorina littorea* has planktonic eggs and larvae which leads to wide gene flow and no genetic isolation. Winkles are virtually uniform in shape and colour. Try a parallel series of measurements for *L. littorea* if you have time.)

Specific questions: a) Are shells more 'squat' on any shore or in any zone of the same shore? Compare shell length (L) and shell width (W).

b) Are shells thicker on any shore or in any zone of the same shore? Compare shell length (L) with lip thickness (T).

c) Is the aperture larger on any shore or in any zone of the same shore? Compare shell length (L) with aperture length (A).

d) Is the body whorl larger (thus the spire shorter) on any shore or in any zone of the same shore? Compare shell length (L) with whorl height (WH).

Methods

On suitable shores select 'upper' and 'lower' zones. If possible find two or more shores of differing exposure. Rank them in increasing wave exposure. Measure the shell dimensions as shown on Fig. 34 for 20-100 individuals at each zone. Small calipers are best for the measurements, and it is sufficient to measure to the nearest millimetre, or perhaps half-millimetre.

NOTE ON CONSERVATION

Please do not damage the shells as you handle them. These measurements can be done on site. **Put each specimen back where it came from – or you are messing-up the gene flow! Dip each one in sea-water and place it back in position. Please do not collect specimens and then throw them back nearby or onto a different shore.** *Nucella* has been decimated on some shores by TBT a chemical from anti-fouling paints and is rare around many ports and harbours. Please do not carry out this project on a shore where there are not many dogwhelks as you might further damage the population.

Results

a) Compile tables of the results from each zone at each site. Keep the various measurements on each individual shell together.

b) Plot scatter diagrams of L against W,T,A,WH for each individual. The gradient of these lines will allow you to compare shell shape and thickness with shore level and exposure. A line fitted by eye is often sufficient to make comparisons. Alternatively, a line can be mathematically fitted using regression analysis (see comments in exercise V).

Discussion

a) Do the shell form ratios vary between shores? Can this be correlated with shore exposure? Do the shell form ratios differ between upper and lower parts of the dogwhelk's range?

b) Suggest reasons for any differences observed? What advantage would a thicker shell confer? What advantages and disadvantages would there be with a larger aperture and hence foot-size? What shell form would be less susceptible to predation? Where would the highest predation pressure be expected?

c) How could you experimentally test some of these ideas?

(Refs: Crothers,1975,1985,1989; Hughes & Elner, 1979; Kitching *et al.*,1966).

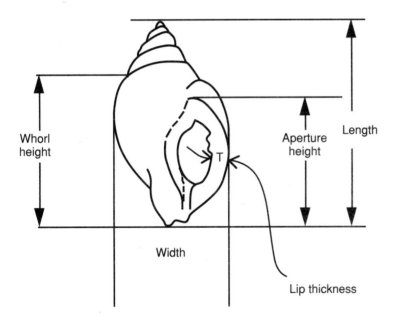

Whorl height

Aperture height

Length

Width

T

Lip thickness

Fig.33. Dimensions to be measured in dogwhelk project.

CLASS EXERCISES

VII. *Fucus vesiculosus* Morphology and Wave Action

Introduction

Fucus vesiculosus is morphologically highly variable: in sheltered conditions it is large, often has many branches and has many bladders; in exposed conditions a short, bladderless form occurs. On shores of intermediate exposure there are various grades in between. These differences could be due to a phenotypic response with environmental control of growth form. Alternatively, there could be an underlying genetic explanation: for example wave action could selectively remove larger, bladdered individuals on more exposed shores; but these forms could be more competitive in shelter. These hypotheses can only be explored by transplant or breeding experiments which are not possible in a short mini-project. However, in a few days the range of morphological variation can be quantified on shores of various exposures. A mixture of bladderless and bladdered adult plants on shores of intermediate exposure would point to an underlying genetic explanation; a uniform gradation in plant size and number of bladders between shores would point to a direct phenotypic growth response.

Methods

This project can be done non-destructively in the field. Alternatively, small numbers of plants can be removed for laboratory measurement. *Fucus vesiculosus* rapidly regrows, so provided only a small number of plants are removed for a **small group** project, little lasting damage will be done to the shore.

a) *Site Choice*: Choose at least 3 sites (5 - 8 are better) on a well defined exposure gradient, such as along a headland or around an island. Rank these sites from most sheltered to most exposed. At each site choose one station in the *middle* of the *Fucus vesiculosus* zone. Characterise the sample area by assessing the cover of total fucoids, *Fucus vesiculosus* and barnacles plus limpet number in 10 random 0.5 m x 0.5 m quadrats.

b) *Plant selection*: Restrict your measurements to adult plants only, that is ones with fruiting bodies at the tips. At each site you need to measure at least 30 adult plants. It is up to the group to decide the exact number to measure.

In each of the above quadrats select the three (or five) largest *Fucus* plants. If you decide to measure more plants, or there are no adult plants in a quadrat throw extra quadrats (in these extra quadrats there is no need to assess the cover etc. as above).

For each plant:
   i) measure its length to the end of its longest axis.
   ii) count the number of bladders on the frond (thallus) along this longest axis (express this as bladders/unit length).
   iii) count the number of branches along this longest axis (express this as branches/unit length).
   iv) if time permits, count the *total number* of dichotomies (branches) in the plant. This gives a measure of bushiness.
      Keep the results for each plant together (e.g. plant 1,2,3 etc.). The largest plants are selected as these represent the oldest fully grown individuals in the population.
   v) Also measure the maximum length of the *smallest* 3 plants showing signs of sexual maturity in each quadrat. This gives a measure of the size of first sexual maturity in a population.

To assess if a mixture of two basic forms occurred on intermediate exposure shores; measure all the adult plants in a quadrat as above, and then repeat until at least 50 are measured.

c) *Measuring exposure*: If the rank order of exposure of the sites is clear and obvious there is no real need to measure exposure. You know that site A is more exposed than B and so on. You could apply Ballantine's scale *but* it largely reflects the balance of fucoids and barnacles from shelter to exposure. Fucoids are used to help determine the scale therefore it is best avoided as the two measures are not independent (discuss this with your group and teacher). You could calculate simple map-based indices of exposure (see p. 90).

Results

a) Work out the mean and standard deviation of fucoid cover, *Fucus vesiculosus* cover etc. for each site.

b) Work out the mean size and its standard deviation for the largest plants at each site.

Both these can be presented graphically as block diagrams for the various shores in ascending order of exposure or shelter. If an exposure measure has been calculated, then this can be plotted on the x-axis to give a line graph. Could you do this with the Ballantine exposure grades?

c) Plot scatter diagrams of number of bladders, number of branches, number of dichotomies against maximum length for each shore. Is there a separation between the various sites? Alternatively work out the mean and standard deviation of bladders/unit length, branches/unit length, and total dichotomies/unit length for each shore and plot as in a) and b).

Discussion

a) Is the change of morphology with increasing exposure gradual or sudden? Are there mixtures of plants of different types on moderately exposed shores?

b) Do you think it is a phenotypic response or the result of selection? What simple experiments could you perform to find this out?

(Refs: Russell & Fielding, 1981; Wright – appendix to Baker & Crothers, 1986).

ENERGY FLOW IN SHORE COMMUNITIES

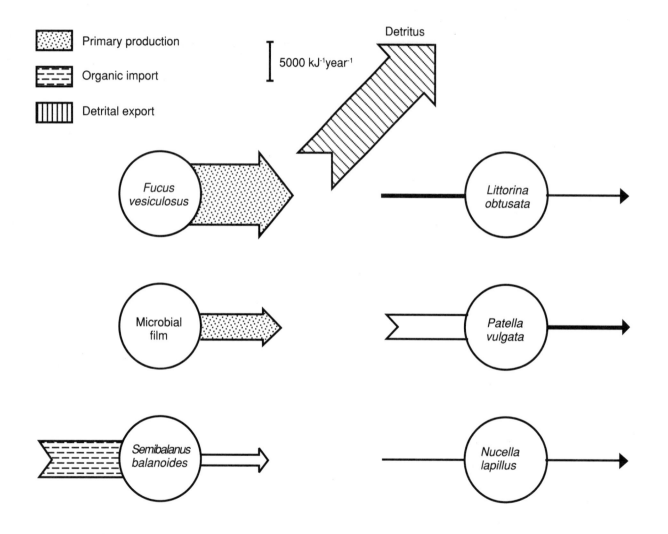

Fig. 34. Energy flow patterns on a moderately exposed shore on the Isle of Man. Production and consumption are indicated by arrows whose widths are in proportion to energy flow. (*Courtesy of R. G. Hartnoll, modified from various sources*).

For a community to function, energy must be either produced within the community or imported from elsewhere. Energy can also be exported from marine communities in various forms including drifting phytoplankton, production of large numbers of propagules or larvae, migration of species using that community as a juvenile or adult feeding ground and probably most importantly as **detritus**.

On British and Irish rocky shores, as elsewhere, there are three main sources of **primary production**. The most obvious is the large attached macroalgae. Less apparent, but of great importance is the microbial film of bacteria, blue-greens, diatoms, fungi, protozoans, and the tiny newly germinated settling stages of macroalgae which invisibly coats the surface of the rock and sometimes makes it very slippery (see Plate VIII, 14 and Plate X). There are also the millions of unicellular phytoplankton which wash over the shore on every tide. A few animals with **symbiotic** algae such as sponges (e.g. *Halichondria*) and anemones (*Anemonia*) occur, but they are nowhere near as important as corals on tropical reefs. A very important source of energy is imported detritus. Subtidal kelps in particular get washed-in, both intact and in fragments, especially during storms and are an important source of food.

Sheltered shores with their large fucoids, and the kelp-covered sublittoral fringe and below of all exposures, are net exporters of energy. Although herbivores are plentiful a relatively small proportion of fucoid and kelp production gets consumed. Most production enters the detritus pathway through continual fragmentation of frond ends, dislodgement of whole plants and the release of dissolved organic matter (DOM) – algae are notoriously leaky. This production is retained within shore ecosystems in various ways. Large pieces of kelp can be caught by sea urchins (*Paracentrotus*), washed-up on the shore where littorinids, small crustaceans and even limpets can consume it, or form large banks of seaweed at the strandline supporting teeming populations of invertebrate scavengers and decomposers. These in turn are often fed upon by large numbers of birds. Further fragmentation and bacterial action both on particulate and dissolved organic matter produces a rich array of particles which can be used by filter feeders.

On exposed and moderately exposed shores the vast numbers of limpets and other grazers feed mainly on the invisible microfloral film. So the energy produced within the community is mainly consumed within it. The film also seems to act as trap for inshore planktonic diatoms, particulate organic matter as well as forming a substrate for precipitation of dissolved organics. These all get deposited when the tide goes out – a bit like the froth left on a beer glass. The filter feeders on exposed shores depend on plankton and small particulates brought past them by currents and waves. Therefore the eulittoral of exposed shores is a net importer of energy. The middle regions of moderately exposed rocky shores are probably modest net exporters: they import less energy than exposed shores and export some fucoids. This will vary with the degree of fucoid cover on the shore (see p. 116). The diagram opposite modified from Hartnoll (1983) tentatively outlines energy flow in the eulittoral of British shores (see Fig. 34).

CAUSES OF DISTRIBUTION PATTERNS

Vertical Zonation

On any environmental gradient, species occur in distinct bands. Going up a mountain, deciduous trees give way to conifers which, in turn, eventually die out, leaving shrubs and herbs, and on the highest peaks only a few lichens can survive. Similarly on the shore, the zones which can occur are broadly related to the stress gradient of increasing exposure to air, and greater fluctuations in conditions, at increasing height above the low water mark.

There have been many laboratory experiments comparing the tolerance to aerial exposure of intertidal plants and animals from different shore levels (see Newell, 1979; Norton,1985 for reviews). Higher shore forms are, not surprisingly, more tolerant. However, if the periods survived are looked at carefully, it is clear that, in many instances, aerial exposure can be tolerated for much longer than it will ever be encountered at the shore levels at which the organisms are found. So, while harsh conditions obviously make high levels of the shore uninhabitable for low level species, what actually sets their upper limit? Conversely, for the lower limits, most intertidal organisms are of marine evolutionary affinity and so it is unlikely that many of them will be adversely affected physiologically by submersion – a marine plant or animal is unlikely to drown! Thus, while the occurrence of zones in themselves point to the underlying effect of the physiological stress gradient, we must be much more careful when trying to decide which factors set the upper and lower limits of a particular species. It is best to consider upper and lower limits separately and to draw the distinction between sessile and mobile animals. Once a plant or sessile animal becomes attached, it can do very little to change its position and hence regulate the conditions it experiences. On the other hand, a mobile animal can move elsewhere to avoid unpleasant conditions: this movement can be to a different zone if dislodged, or a short distance to a more favourable microhabitat.

Upper limits of plants and sessile animals

There is considerable evidence from direct observations that upper limits can be set by physical factors directly killing the adults or young stages of plants and sessile animals. Young germlings, plus the larvae and juveniles of animals, are usually more vulnerable than adults, so upper limits may often be set at, or soon after, settlement.

On sheltered shores, *Pelvetia, Fucus spiralis* and, more rarely, *Ascophyllum* all get bleached and die during periods of hot weather (e.g. Schonbeck & Norton, 1978; Hawkins & Hartnoll, 1985; Norton,1985). Fucoids are particularly vulnerable to sudden changes in temperature and relative humidity as they have some ability to increase tolerance in response to gradual sub-lethal change (Schonbeck & Norton, 1979c). On moderately exposed shores, *Laminaria digitata* and many low shore red seaweeds, particularly the delicate species usually found in the understorey, are also vulnerable to spells of both hot weather and extreme cold (Todd & Lewis, 1984; Hawkins & Hartnoll,1985).

On more exposed shores, mussels grow more slowly at upper levels, presumably due to desiccation stress coupled with reduced feeding time, so it is likely that their upper limits are directly set by physical extremes (Seed, 1969). Juvenile *Semibalanus balanoides* settling above the adult zone will be killed during a hot spring (Connell, 1961a). In a cool spring, a slight range extension can occur because if the juvenile phase is survived, the adults are less vulnerable. Every so often a very hot summer will kill off these adults, or they will eventually die and not be replaced by new settlement.

Various experimental studies have also provided evidence that upper limits are set directly by physical factors including field transplant experiments, artificially making the rock wetter by run-off, and laboratory cultures and simulations (e.g. with barnacles, Foster, 1971; fucoids, Schonbeck & Norton, 1978). Evidence has also come from observations of death following uplift of the whole shore following earthquakes and atom bomb tests, thankfully not in the British Isles.

Until recently, it was firmly stated in many text books that upper limits are set directly by physical factors. This is certainly the case for many upper and mid-shore species. However, some mid- and low-shore species (e.g. *Fucus vesiculosus*, *F. serratus*) have not been shown to die at their upper limit, even during extremely hot weather when some high shore species are getting badly affected (Schonbeck & Norton, 1978; Hawkins & Hartnoll, 1985). In the absence of physical factors, we must look for a biological explanation.

On sheltered shores, experimental removal of *Fucus vesiculosus* above the *F. serratus* zone and similarly *F. spiralis* removal from above the *F. vesiculosus* zone can lead to a slight extension of the lower species up the shore. *Fucus serratus* will also grow upshore in the *Ascophyllum* zone when the latter is removed (Hawkins & Hartnoll, 1985) and *F.serratus* survive when transplanted slightly above their normal zones (Schonbeck & Norton,1978). A few *Laminaria* sporelings can grow in the absence of experimentally removed *Fucus serratus* (Hawkins & Hartnoll, 1985) but will be killed by hot weather in some years. These experiments all suggest that competition is restricting the *immediate* upward extension of these species.

On more exposed shores, if the grazing limpets are removed from above the low-shore red algal turf, then this turf will extend higher up the shore and survive subsequent summers (see Plate VIII, 15). When thousands of limpets were killed by detergents used to clean up the oil spilled by the wreck of the "Torrey Canyon", reduced grazing pressure led to upshore extension of the low shore reds and kelps (Southward & Southward, 1978). *Fucus serratus* can also be induced to grow high up moderately exposed shores if limpets are removed. Therefore grazing can directly limit the upward extension of primarily low-shore species in the British Isles. Results of this kind were first clearly shown in Australia (e.g.Underwood, 1979; Underwood, 1980; Underwood & Jernakoff, 1981).

On moderately exposed shores, where fucoid cover is patchy and discontinuous, competition is unlikely to cause the diffuse zonation patterns observed. Here, the ability to grow quickly and reach a size (> 5 cm) immune from limpet grazing is crucial for survival. Species adapted to a particular zone would be expected to grow faster and, on average, escape grazing more frequently (see Hawkins & Hartnoll, 1983b for review). For example, *F. serratus* grows faster when cultured under low shore conditions than mid-shore *F. vesiculosus* and high-shore *F. spiralis*, and several species of fucoids grow more slowly at the top of their zone than lower down within it (Schonbeck & Norton, 1980a). Consequently, occasional patches of *F. spiralis* and *F. serratus* can be found in the *F. vesiculosus* zone – the young stages do manage to escape grazing locally, although infrequently.

When limpets are removed, or when new breakwaters are colonized, considerable jumbling up of the fucoid zones can occur at first – until competition and grazing sort and focus the zonation patterns.

Lower limits of plants and sessile animals

Some intertidal plants and animals will grow subtidally if space is made available for them during the settlement season (e.g. ephemeral algae and the barnacle *Semibalanus balanoides*). It is likely that many plants and animals would similarly thrive lower down the shore if given the chance. Biological factors, such as competition, grazing and predation, are most likely to prevent downwards extension.

Fucus spiralis transplanted on rocks lower down the shore grows better than control transplants within its own zone (Schonbeck & Norton, 1980a). On sheltered shores, the lower shore plants are generally bigger and bushier than the upper shore ones and grow much faster. This makes them more competitive. If *Fucus spiralis* is removed just beneath the *Pelvetia* zone, *Pelvetia* can be induced to extend downshore (Schonbeck and Norton, 1980a). Similarly, *F. spiralis* can be induced into the *F. vesiculosus* zone, *Fucus vesiculosus* into the *Ascophyllum* zone and *Himanthalia* into the *Laminaria* zone (Hawkins & Hartnoll, 1985). Similar work in New England has shown that *Fucus distichus* is prevented from extending lower down by *Chondrus* (Lubchenco,1980).

The mid- to low-shore barnacle, *Semibalanus balanoides*, has been shown to be competitively superior to *Chthamalus montagui* in experiments conducted in the Firth of Clyde (Connell, 1961b). Removal of *S. balanoides* from stones bearing *C. montagui* transplanted into the *S. balanoides* zone, resulted in enhanced survival of *Chthamalus*. On untouched areas, *C. montagui* was ousted by the more rapidly growing *S. balanoides*. In work done on the eastern seaboard of the USA, *Mytilus edulis* has been shown to outcompete *S. balanoides*, particularly on horizontal surfaces (Menge, 1976).

Grazing can also prevent downshore extension of many species. *Fucus spiralis* occurred lower on the shore in experiments in which all the limpets were removed from a 10 m wide strip down the shore in the Isle of Man. Many kinds of ephemeral algae, usually found in the littoral fringe (e.g. *Porphyra, Enteromorpha, Blidingia*), will occur in grazer removed areas in the eulittoral (see Hawkins & Hartnoll, 1983b for review). *Semibalanus balanoides* are thought to have their lower limit set by predation by dogwhelks (Connell, 1961b), although competition with red algal turfs and the increased sweeping effects of low-shore fucoids, which can dislodge settling cyprids, are also important (Hawkins, 1983). On the Pacific coast of the United States, predation by large starfish has been shown to be responsible for setting the lower limit of *Mytilus* (Paine, 1974). It is possible that our own starfish, *Asterias*, can have a similar effect on some shores.

The direct effect of physical factors cannot, however, be completely discounted. The high-shore alga *Pelvetia* degenerates and eventually dies when transplanted lower on the shore (Schonbeck & Norton, 1980a; Rugg & Norton, 1987). *Chthamalus montagui* also die more rapidly lower on the shore; newly metamorphosed juveniles seem particularly vulnerable to too much submersion in experiments conducted with an artificial tidal aquarium (Burrows, 1988).

Larval behaviour and zonation

Generally, the propagules of plants have little control over where they settle. Some species, such as the fucoids, have heavy eggs or zygotes which limit dispersal. Other species (e.g. *Enteromorpha*) have motile zoospores, which can respond either positively or negatively to light thus having some control over settlement and attachment position, although the longevity and strength of swimming is limited.

In contrast, many larvae of sessile marine invertebrates have considerable swimming powers, and the capacity to delay metamorphosis allowing considerable choice of site of settlement (see Crisp, 1974; Newell, 1979 for review). Many of the settlement cues ensure that the juveniles settle in the same zone as the adults. The larvae of most intertidal species are positively phototactic, ensuring a high position in the water column and hence a greater chance of encountering the intertidal. Barnacle cyprid larvae are attracted by pits and grooves on the rock surface and also by chemicals present in the adults. Mussel larvae like fibrous textures, similar to the dense byssus threads by which clumps of adults adhere to rocks. Spirorbids which live on seaweeds are attracted to settle by chemicals exuded by a particular species of host plant. These mechanisms not only increase the chances of survival but also the likelihood of successful reproduction, whether by external fertilization (e.g. sponges, mussels, limpets, tubeworms) or internal fertilization (e.g. barnacles, winkles).

So, zonation patterns are maintained by the larval behaviour patterns of many species. These increase the likelihood of survival in the particular physical environment and may reduce the risks of encountering harmful biological interactions such as competition or predation. Late-arriving larvae, finding suitable sites already occupied, become less fussy as their energy reserves run out. Many of these die, but some are able to colonize new free space, such as recently built breakwaters, piers and ships bottoms.

Mobile animals

High on the shore, mobile animals are sometimes killed by unfavourable conditions, so setting upper limits. This particularly occurs when a species borders an unexploited resource (Wolcott's Hypothesis, 1973). For example, high shore limpets often live just below a luxurious winter lawn of blue-green algae, diatoms and ephemeral green algae. All this food acts as a temptation and they "push their luck" by extending their range upshore. As a result, during spring, limpets often die, caught out by rapidly changing conditions.

At the bottom of the shore, predation by crabs, fish and starfish can be responsible for setting the lower limits of mobile animals, although the evidence is not clear. The lower limit of British limpets may be set by competition for space with rapidly growing algal turfs. At a certain level, the seaweeds probably grow faster than the limpets can graze, thus setting their lower limit .

Most of the time, however, behavioural patterns are directly responsible for setting the distribution patterns of mobile animals. They behave in various ways to stay within the set of conditions (or zone) in which they can tolerate the physical environment and reduce the risk of biological interactions – some insidious, like competition, others terminal, like predation!

When dislodged, shore gastropods instinctively crawl upwards. Wave dislodgement tends to sweep animals downshore, so this behaviour helps animals regain their appropriate shore level. Many species (e.g. *Littorina obtusata* and *L. mariae*, Williams, 1987) have quite refined homing ability and will return to their own shore level if displaced either up or down the shore. After feeding excursions, animals often return to a home area. In dogwhelks this may be a crevice and in limpets a closely defined home scar, which the shell fits exactly. Shore animals also display rhythmic behaviour ensuring that they are active during favourable conditions. Shore crabs (*Carcinus maenas*) are active at night and at high tide (Naylor, 1985 for review). Limpets forage when submerged during the day and also in humid conditions when the tide is out, during both the day and the night, but they do not seem to forage when submerged at night, perhaps to avoid predators (Hawkins & Hartnoll, 1983b; Little, 1989). *Littorina saxatilis* is primarily active during humid tide-out conditions, thereby avoiding dislodgement by waves when the tide is in.

Horizontal distributions

Very little experimental work has been done on the respective roles of physical factors and biological interactions in setting the distribution patterns of littoral fringe species. Greater wetting in exposure is obviously responsible for the greater height of the littoral fringe and probably directly causes the greater abundance of ephemeral algae (e.g. *Porphyra*). They are less likely to die off in the summer and persist for longer periods of the year in exposed conditions. The respective roles of physical factors (such as greater risk of dislodgement in exposure and greater risk of desiccation in shelter) and biological factors (competition, predation) have not been investigated much on the high shore. There is plenty of scope for investigation, particularly with the high shore winkles: for example is *Melaraphe neritoides* better suited to exposed conditions? What restricts it from occurring on sheltered shores? Is competition with *L. saxatilis* important or is it larval supply?

Most work has been done in the eulittoral. Early explanations emphasized the direct effects of wave action in preventing fucoids from extending into exposure. This certainly seems the case with *Ascophyllum*. However, some pioneering work on the Isle of Man in the 1940s (Jones, 1948; Lodge, 1948) showed that physical factors were not responsible for preventing growth of *Fucus vesiculosus* on moderately-exposed barnacle dominated shores. Limpets were removed from a 10 m wide strip down the shore and within 2 years the shore was dominated by a luxurious growth of fucoids. Plate IX, 1-3 show a recreation of this experiment on a smaller scale. Clearly it is the direct action of the grazers which prevents growth of *Fucus*. Recent work (Plate IX, 4) has confirmed the importance of limpet grazing (see Hawkins, 1981a,b; Hawkins & Hartnoll, 1983b); limpets were removed from an area of shore and fences prevented their re-entry. A dense sward of *Fucus* occurred within 9 months. The limpets were removing the microscopic early stages of the fucoids as they settle and germinate on the rock. Even on the most exposed shores grazing can be important. Many limpets were killed by application of dispersants to clean up the Torrey Canyon oil spill in West Cornwall in 1967. Fucoids managed to grow on even exposed shores such as Sennen Cove on the Atlantic Coast of Cornwall (Southward & Southward, 1978).

Although limpets prevent settlement and germination, wave action probably restricts survival of plants, particularly if they are growing on barnacles which do not provide a secure anchorage. So really we have a dynamic equilibrium – a bit like some chemical reactions. On wave-beaten shores conditions are favourable for limpets, there are many of them and they prevent fucoid growth which allows barnacles to grow. Even if a *Fucus* plant manages to escape grazing it is unlikely to survive for long – although a stunted bladderless form of *Fucus vesiculosus* does occur on more exposed shores. In shelter the balance is tilted the fucoids way. They can grow and survive well. Limpets are rare in shelter. The exact reasons are unknown but silting, lack of larval input of both barnacles and limpets due to barrier and sweeping effects of the seaweeds and lack of 'lithothamnia' pools which limpets may like to settle in could all contribute. Again this area needs further work. Fig. 35 is a diagrammatic representation of these ideas. There is some evidence that at more northerly latitudes conditions are more favourable for fucoids which extend out more into exposure. Whereas further south with more species of grazers the balance is tilted towards limpets and barnacles which extend further into shelter. Inshore changes of climate with latitude seem to shift the fulcrum or balance point.

Low on the shore there is a similar interplay of physical factors and biological interactions. *Laminaria digitata* dominates on moderately exposed shores but gives way to *Alaria* in exposure and *L. saccharina* in shelter. If *Laminaria digitata* is removed on a moderately sheltered shore *Alaria* and *Laminaria saccharina* can be induced to grow side by side. Both are opportunist species and are out-competed in moderate exposure by *L. digitata*. In exposed conditions storms make gaps amongst the *L. digitata* or completely denude it allowing *Alaria,* with its better attachment and elastic

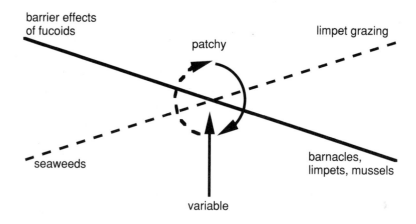

On moderately exposed shores cycles due to tipping of balance by environmental factors acting through recruitment

Fig. 35. A simple explanation of the effects of wave action on the balance between fucoids and limpets plus barnacles on rocky shores (*modified from Hawkins and Hartnoll, 1983b*).

frond, to thrive. In sheltered conditions unstable boulders favour the annual *L. saccharina* because winter disturbance prevents dominance by *L. digitata* (Hawkins & Harkin, 1985).

Summary

In sessile organisms, physical factors directly set the upper limit of most upper and some mid-shore species. Some low-shore species, such as *Laminaria digitata*, may have their upper limits set for some of the time by competition, with occasional physical extremes also being important. Many mid- to low-shore species have their immediate upper limits set primarily by biological factors, such as competition and grazing. Indefinite extension up the shore, in the absence of biological checks, is prevented, however, by adverse physical conditions. The greater competitive ability of the species in the zone above, or increased grazing pressure, reflects an underlying change in the physical environment. Lower limits are generally set by competition, grazing and predation, with the exception of *Pelvetia*.

In sessile animals, selective settlement behaviour reduces waste of reproductive effort by, to some extent, predicting conditions in which the adult will survive. The distribution of mobile animals is directly set, primarily by their behaviour patterns. These have evolved as a response to avoid unfavourable physical conditions (usually upshore) and unfavourable biological interactions (downshore). In these mobile animals, physical extremes on the high-shore can be important in setting upper limits; competition and predation can set lower limits, but there are few clear examples.

On the horizontal exposure gradient the respective roles of physical factors and biological interactions are less clear. Both physical factors such as wave action and biological factors such as grazing and competition can prevent organisms extending into exposed conditions. Similarly physical factors, such as silting, and biological factors such as predation and competition, can prevent organisms extending into sheltered areas.

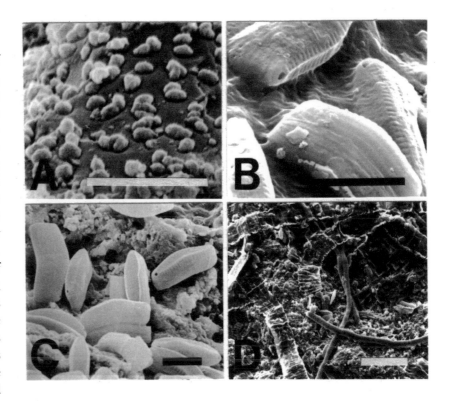

Plate X. The microbial film which coats the rock surface on shores. A) Bacteria colonizing newly settled detritus (bar = 4 µm) B) The diatom *Achnanthes* incorporated into the slime film (bar as above). C) *Achnanthes* (bar = 10 µm). D) *Fragilaria* (bar = 100 µm). (*photograph, courtesy of Dr Andrew Hill*).

DYNAMICS OF SHORE COMMUNITIES

Seasonal Changes

The composition of the shore community varies with the seasons. When light increases in spring, many ephemeral algae proliferate, particularly *Enteromorpha* and *Ulva*, giving the shore a green appearance. In the littoral fringe many species of algae, such as *Prasiola*, *Porphyra* and a brown slippery slime (composed of small unicellular plants such as diatoms and blue-greens) proliferate over the winter. These may well be present on an Easter vacation fieldcourse, but will have dried out, died and disappeared by May or June, except in the most dismal of British summers. A September field course would therefore not see them.

There are also changes due to the pronounced seasonality of reproduction of many littoral animals in temperate waters. The barnacle, *Semibalanus balanoides*, settles in late winter/spring (February-June, depending on locality). If settlement is heavy, the shore will be covered by little (1 mm) sausage-shaped cyprids, the stage that settles from the plankton (see Plate VI, 2). These rapidly metamorphose into pink and then white mini-adult barnacles and start to grow rapidly. *Chthamalus* settle in the summer and autumn but have much smaller cyprids.

Activity patterns change with season. More crabs are found on the shore in the summer. In spring and summer dogwhelks actively forage, but in the winter they tend to occupy cracks and crevices. Winkles and topshells behave similarly, and they also they tend to occupy higher shore levels in the summer.

Long-term Changes

Plate IX, 5 & 6 shows changes on a Manx shore over a period of several years. You can see that the shore has changed from barnacle domination to being covered with seaweeds. Are all shores this variable? What generates change of this kind?

In answer to the first question, many shores are exceedingly stable and remain much the same for many years. Generally, the vertical zonation pattern remains fairly constant, although the composition of a particular zone may change. Exposed shores always tend to be dominated by mussels and/or barnacles. On shores with mussels, considerable fluctuations in their abundance can occur: sheets of mussels get dislodged by storms and settlement of larvae from the plankton can be very variable. Barnacles then take the place of the mussels. Sheltered shores also tend to vary little, being always covered by seaweeds arranged in clear zones. The boundaries of the zones may shift up and down a little bit, but their relative positions will not change. In contrast, the eulittoral of shores of intermediate exposure tends to be much more variable. Patches of fucoids, barnacles and bare rock occur, interspersed with limpets, and these patches change with time (Baxter *et al.*, 1985; Hartnoll & Hawkins, 1985a).

To answer the second question we have to consider the balance between domination by filter-feeders and grazers in exposure and seaweeds in shelter (see previous section). This is delicately poised on moderately exposed shores. Fluctuations in the recruitment of the various species (e.g. Bowman & Lewis, 1977; Hawkins & Hartnoll, 1982a; Kendall *et al.*, 1985) can tilt the equilibrium one way or another: a reduction in limpets can reduce grazing pressure, allowing seaweeds to flourish; good conditions for fucoid recruitment can swamp the grazing ability of limpets; an increase in limpets can reduce seaweeds; good barnacle settlement can lead to dense patches of barnacles, amongst which limpets cannot easily graze, ultimately leading to seaweeds escaping grazing and proliferating (see Baxter *et al.*, 1985; Hawkins, 1981a, b, 1983; Hawkins & Hartnoll, 1982a, b, 1983a, b; Hartnoll & Hawkins, 1985a). Fig. 36 gives some idea of the changes occurring in a 2 m by 1 m area on the middle of a moderately exposed shore. Fig. 37a, b outlines the sequence of events occurring on patchy shores when a clump of seaweed grows in an area of low grazing pressure. There is a basic cycle which to some extent is self-perpetuating: the aggregation of limpets in one area results in low grazing pressure elsewhere which allows the cycle to restart in an

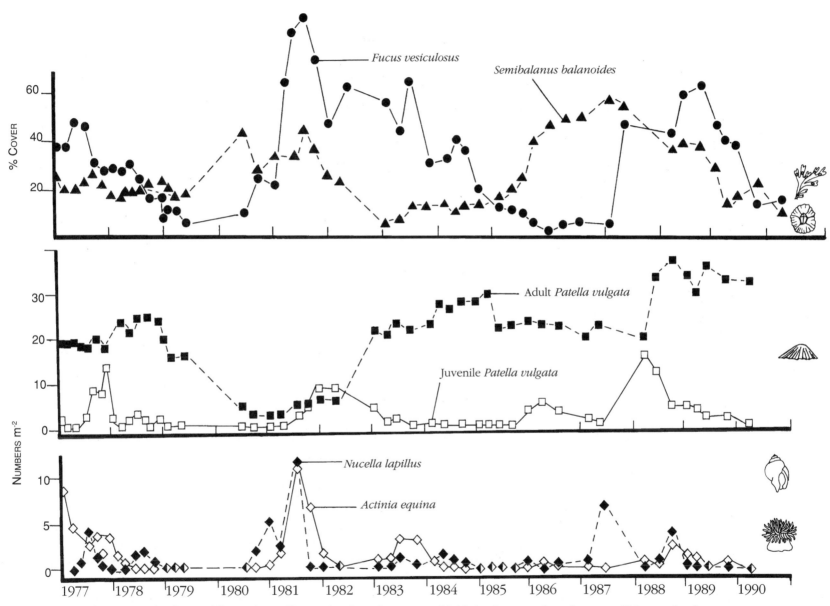

Fig. 36. An example of natural fluctuations. Changes in a 2m x 1m area at mid tide level on a moderately exposed Manx rocky shore.

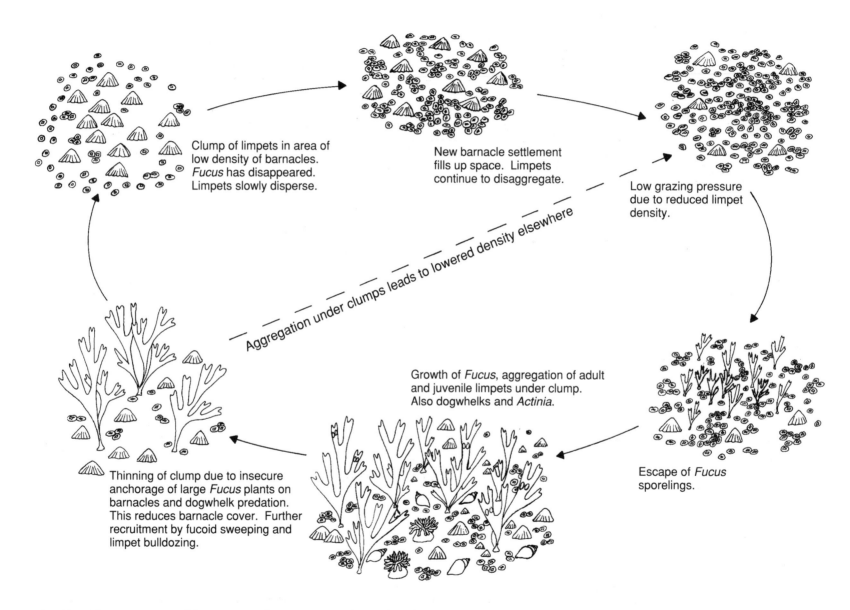

Clump of limpets in area of low density of barnacles. *Fucus* has disappeared. Limpets slowly disperse.

New barnacle settlement fills up space. Limpets continue to disaggregate.

Low grazing pressure due to reduced limpet density.

Aggregation under clumps leads to lowered density elsewhere

Growth of *Fucus*, aggregation of adult and juvenile limpets under clump. Also dogwhelks and *Actinia*.

Escape of *Fucus* sporelings.

Thinning of clump due to insecure anchorage of large *Fucus* plants on barnacles and dogwhelk predation. This reduces barnacle cover. Further recruitment by fucoid sweeping and limpet bulldozing.

Fig. 37a. Pictorial representation of the sequence of events on a patchy moderately exposed shore over several years, based on the shore studied in Fig. 36.

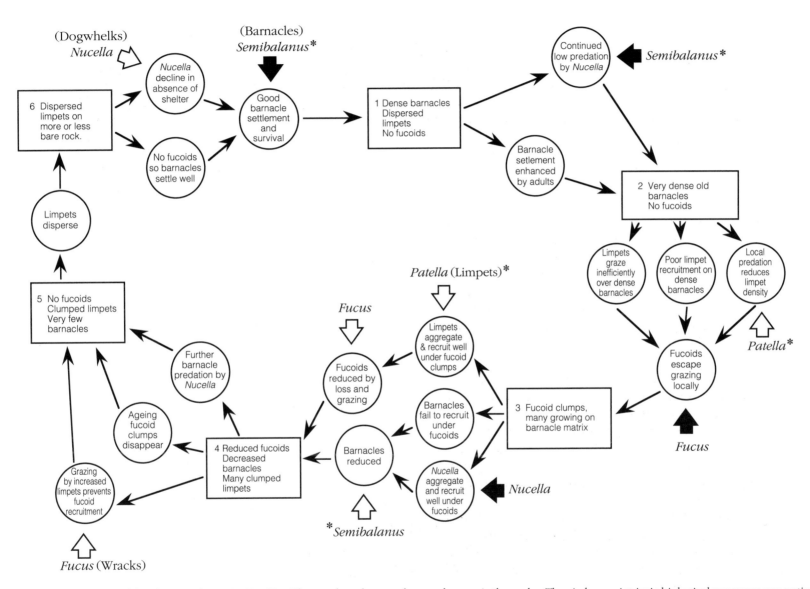

Fig. 37b. A flow chart of the changes shown in Fig. 37a. The numbered rectangles are changes in the cycle. The circles are intrinsic biological processes generating and maintaining the cycle. The heavy arrows are where very good settlement of the named species either promotes (solid arrows) or inhibits (open arrows) the progress of the cycle. The asterisks indicate where settlement is from a highly variable planktonic phase.

DYNAMICS OF SHORE COMMUNITIES

adjacent patch. Irregular predation by seabirds can also create localized areas of low limpet density allowing algae to grow. The effects of variable recruitment on the cycle in either speeding it up or slowing it down are shown (Fig. 37b). So, remember that any transect is a frozen frame in time: on some shores it may only represent one facet of a constantly changing community; on others the shore may always be much the same.

Overview

The diagram opposite (Fig. 38) provides an overview of the processes structuring rocky shore communities in the North-east Atlantic. It applies to shores of moderate wave exposure or greater, and provides an explanation of the Stephenson's "universal scheme of zonation" (see p. 13). The community in the littoral fringe is structured primarily by the stressful, highly variable and unpredictable physical environment. Biological interactions such as grazing and competition can be important here but the physical regime is of overriding importance. Most algal species are ephemeral and seasonal in occurrence and animals are confined to cracks and crevices.

The eulittoral zone is dominated spatially by barnacles or mussels but is strongly structured by grazing of *Patella* and other gastropods. Predation by *Nucella* may also be important in the low eulittoral. The barnacles are only present by virtue of the limpets keeping competing algae in check.

In the sublittoral fringe algal growth exceeds the capability of limpets and other gastropods to control it. This has been shown best by work in Australia, but recent work in the UK (see Plate VIII, 15) suggests it is the case here as well. The ability of algal turfs to grow and successfully compete for space, and the canopy effects of kelps, are the dominant interactions structuring the communities at this shore level. Conditions are optimal for algae at, or a little way beneath, the low water mark where light is sufficient, submersion is continual and nutrients are plentiful. Strong water movement in this area probably inhibits grazers, particularly sea urchins.

This analysis can be extended into the sublittoral. With increasing depth, conditions become less suitable for algae as light diminishes. Consequently, the canopy effects become less important in structuring the algal community. As water movement decreases, grazing by sea urchins can become predominant and can set the lower limit of *Laminaria* (Kain, 1979). A more open ('park') area of very small reds is found below the *Laminaria* zone. At yet greater depths a physical factor, availability of light, re-asserts its control on algae. Conditions continue to be favourable for animals, however, and their greater competitiveness may amplify the effects of diminished light.

DOMINANT STRUCTURING AGENCY

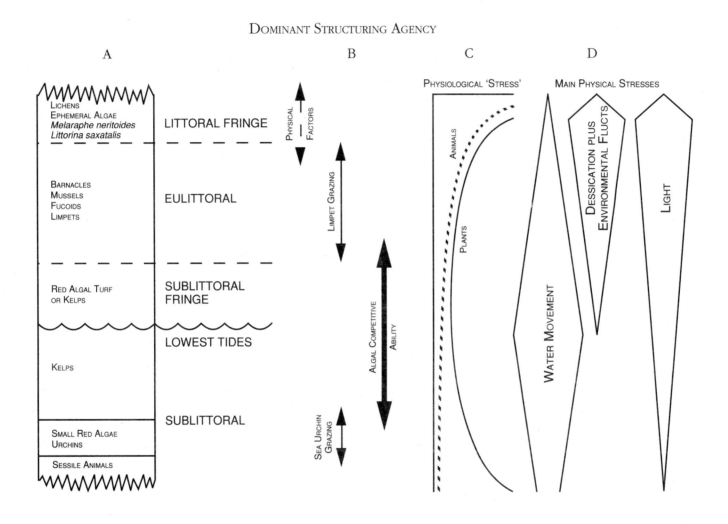

Fig. 38. An overview of processes structuring the intertidal and shallow subtidal of shores of above moderate exposure. It also serves as an explanation of the Universal scheme of the Stephensons. A: The pattern. B: Dominant structuring agencies are both physical and biological, overlap suggests a balance between factors. C: Generalized physiological stress gradients. D: Main physical stresses causing physiological stress for marine algae and animals. (*Modified from Hawkins and Hartnoll, 1983b*).

PROJECT SUGGESTIONS

To save space the following are outlines of projects suitable for either single students or small groups (<10) which work well in a period of 2-7 days. In many cases they can be amplified to longer term projects for 'A' levels etc. One or two key references are given in each case to help in the planning and organisation of the project.

Field Projects

Age and growth in *Monodonta*

Monodonta lineata can be aged quite easily by rings on the shell which allows the age as well as length structure of populations, and the growth rate to be assessed. Comparisons can be made between the shores in the same area, between shore levels and even between widely separated shores (refs: Williamson & Kendall, 1981; Kendall,1988).

Activity patterns of intertidal gastropods

When do winkles, topshells, dogwhelks and limpets forage? By observing populations of intertidal gastropods in a set area the proportion active (% activity) can be monitored over 24-96 hours. In the case of littorinids count the number actively crawling out of the total number in an area (50-100 individuals will give good results). This project can be done with high shore littorinids on neap tides without getting wet. Simple snorkelling gear or scuba gear allows observations at high tide. In very sheltered or calm conditions a viewing bucket can be used from the shore or a boat. By marking a stripe down from the apex of the shell of the limpet and over to the rock of the home scar it is easy to assess whether a limpet is active or not by judging if it is present or absent from its home scar. Comparisons can be made between species with overlapping distributions (e.g. *Melaraphe neritoides* and *Littorina saxatilis*) (refs: Hawkins & Hartnoll, 1983b; Little 1989; Little *et al*,1988 for further references and experimental details).

Relationship between prey size and predator size using *Nucella*

In any area with dense *Nucella* populations, some will be found feeding when the tide is out: they will have their proboscis inserted inside barnacles, or be drilling holes in mussels or more rarely littorinids, and sitting on top of small limpets. Work out the proportion of different sized dogwhelks feeding at different shore levels and what their prey is. Measure the size of the prey (aperture size in barnacles) and the size of the dogwhelk feeding on it. Plot appropriate scatter diagrams. Is prey size related to predator size? This can be formally tested using correlation analysis (refs: Hughes & Dunkin, 1984a, b; Crothers, 1985 who also suggests many other projects which can be done using dogwhelks).

Barnacle population structure

The population structure of intertidal barnacles changes with shore level, wave exposure and microhabitats. Comparisons can be made for single species (e.g. *Semibalanus balanoides* in the north) or several species (e.g. *Chthamalus stellatus*, *C. montagui* and *Semibalanus balanoides* in the south and west) depending on locality. Physical factors such as desiccation are likely to be important at the top of the species range leading to a population structure dominated by old barnacles, with few juveniles due to infrequent good settlement years. Low on the shore only juveniles may survive due to size-selective predation. This project is particularly good in April, May, June when newly settled *Semibalanus* can be examined (refs: Connell, 1961a,b; Hawkins & Hartnoll, 1982a; Hawkins, 1983; Rainbow,1985).

Distribution of *Actinia fragacea* and *Actinia equina* colour morphs with and between shores.

Recent work has suggested that some of the different colour morphs of *Actinia* are different species. There is also some evidence that there are habitat differences, particularly between the greens (*Actinia prasina*), the strawberries (*Actinia fragacea*) and the other red/brown forms of *Actinia*. The distribution in terms of zonation with wave exposure and microhabitat differences (pools, crevices, understones) can all be easily investigated. (Carter & Thorpe, 1981; Quicke *et al.*,1983; Haylor *et al.*,1984; Quicke & Brace,1984; Sole-Cava & Thorpe,1987).

Bad Weather Projects

The following projects can be done in primitive laboratory facilities using simple apparatus. They are included because although the authors feel that wherever possible work should be done in the field, there are some days on some field courses where indoor reserves are very useful! In all these projects a quick collection of material on the shore, followed by laboratory work can lead to very interesting results.

Food preferences of topshells and littorinids

Littorinids and topshells are very amenable experimental animals, although it is helpful to starve animals for a day or two before doing feeding experiments. Some algae are preferred to others by grazers. Many such as fucoids have unpleasant chemicals (polyphenols) which are used as a defence, others such as *Corallina* have structural defences such as chalk inclusions, whilst others are just difficult to handle.

Littorina littorea is probably the easiest species to use and groups of snails (usually 30 or so) can be presented with different species of seaweeds in pairs and their preferences assessed. Another way of testing preferences is to give a group of winkles in the middle of a tank the choice of several seaweeds placed in a ring around them. Shallow tanks with a shallow layer of water are best as *Littorina* tend to climb tank walls when disturbed. It is a good idea to cover the tanks to get rid of any confounding directional cues such as light intensity. Tanks can be turned around to check for any inherent directionality due to their construction or location. Alternatively, groups of snails can be given set amounts of seaweeds to eat and the rate of consumption in terms of changes of wet weight assessed. Try comparing fucoids with various reds and greens. Seaweed extracts made by crushing them in water can be impregnated into filter paper and the effect of chemical cues in the absence of textural cues assessed.

In *Littorina obtusata* (found on *Ascophyllum*) and *Littorina mariae* (found on *Fucus serratus*) movement towards seaweeds is probably due mainly to position maintenance behaviour as the alga is both their home and food source. Do these species always go towards their host seaweed? If given no choice how do rates of consumption of *L. obtusata* compare on *Ascophyllum* with other species of fucoids plus reds and greens (do not use *L. mariae* as it feeds mainly on epiphytes) (refs: Bertness *et al.*,1983; Watson & Norton,1985a,b, 1987; Imrie *et al.*, 1989).
Always return animals to their shore after use.

Tolerance of barnacles to desiccation

Barnacles can be transported easily to the laboratory by chipping off small pieces of rock with 50 or so individuals on. These can then be placed in various conditions of humidity and temperature using tanks and such simple expedients as soaking paper towels in seawater and keeping them in covered or uncovered tanks. Relative humidity can be measured with an hygrometer. It is easy to distinguish the dead barnacles, as the opercular plates covering the feeding apparatus collapse leaving a black gaping hole. If the barnacle has just died, however, it may be necessary to touch the opening between the opercular plates with a needle to detect lack of movement. Comparisons of mortality can be made either between species collected at the same shore level, or at the top of their range or within species or populations taken from different shore levels or shores of different exposures. Specimens from high on sheltered shores would be expected to be most tolerant (why?) (refs: Barnes & Barnes, 1964; Foster 1971; Newell,1979).

The effects of desiccation on intertidal fucoids and other seaweeds

To compare the degree of adaptation to stressful emersion by intertidal seaweeds, the rate of water loss can be compared (by weighing) when placed in a low humidity/high temperate environment. Good species to compare are *Pelvetia, Fucus spiralis, F. vesiculosus, F. serratus* and *Ascophyllum.* The comparison can be extended to others particularly kelps, reds and greens from various shore levels. All that is needed is a reasonably accurate balance and lots of dishes. It is also instructive to make comparisons of rates of loss over periods that simulate the time that each species is emersed in the field. The results will be surprising! (refs: Jones & Norton,1979,1981; Schonbeck & Norton,1979a,b; Norton 1985).

Escape responses of limpets to starfish

Patella vulgata reacts strongly to starfish such as *Asterias.* Small limpets run away, large limpets raise their shell, stick out their pallial tentacles, 'mushroom' and 'stomp' on the arms and tubefeet of the starfish, often driving them away. The key to this project is to use limpets which have not been badly disturbed. To do this, remove small boulders with limpets on them at low tide. Gently take these boulders to the laboratory and add fresh, well aerated seawater to the tank to simulate the incoming tide. Use one size of starfish to test limpets of different sizes. Hold it next to the limpet when testing. What is the size at which the limpets run or stand and fight? Try different sizes of starfish. What does the starfish do if not held next to the limpet? Try adding to the tank an extract made by washing a starfish in a small volume of seawater and filtering it. Do they still respond? The extract can be put on a cotton bud (use one soaked in seawater as a control) to simulate the starfish. Is the response due to chemicals in the water (chemoreception) or is it due to contact plus chemical cues (contact chemoreception)? Comparisons can be made between species of *Patella* or with *Acmaea* and *Helcion.* Other gastropods such as topshells can be tried (ref: Branch, 1979,1981).

NOTES

Monitoring

Field courses which return year after year to the same place can build a good picture of the dynamics or the stability of shore communities – provided methods and identification are consistent. If as an individual you live and work near the sea there is also considerable scope to build up information on changes on long and short time scales by repeatedly visiting a chosen site.

One of the simplest methods is to take photographs of the whole shore (see Plate IX) from a set vantage point. If required, selected smaller areas at different tidal levels or exposures can be photographed. A wide-angle lens will allow areas of up to 10 m x 10 m to be photographed from quite short distances. Smaller areas, such as 1 m x 1 m or 0.5 m x 0.5 m, quadrats can also be photographed from standing with a wide-angle or semi-wide lens. Using a macro lens, areas as small as 0.05 m x 0.05 m can be monitored for changes in barnacle or spirorbid populations, although a tripod and a flash usually helps. Very interesting series of photographs can be obtained which can be studied and measured in the warmth of the laboratory.

More detailed studies can be made by non-destructively sampling a marked transect, or fixed areas at set levels on the shore. Alternatively a selected area can be repeatedly sampled using suitable numbers (10-20) of random quadrats (0.5 m x 0.5 m). If adopting the fixed quadrat approach then you should select an area which can be sampled without treading inside the quadrat – 2 x 1 m or 4 x 1 m are quite good for this (see Lewis,1976; Hawkins & Hartnoll,1983b; Hiscock,1985; Baker & Crothers,1986). From a statistical viewpoint, multiple random quadrats are probably the best method for monitoring.

Population structure of littorinids, topshells, limpets and dogwhelks can be monitored as well. The MCS is promoting a scheme to monitor populations of dogwhelks. Many populations near harbours and marinas have sufferred from the effects of leachates from toxic antifouling paints which cause females to take on male characteristics by growing a penis ("imposex") and eventually become sterilized causing a decline in populations (Bryan *et al.*, 1987; Gibbs *et al.*,1987). Now these paints have been banned it is an ideal opportunity to monitor the recovery of populations (contact MCS for details).

Another long-term, but very simple, behavioural experiment is to mark the positions of 30 or so limpets at a particular shore level. This is easily done by marking three reference points and measuring the distance to each. The position of the limpet can then be determined by triangulation (see Hartnoll & Wright,1977). By visiting the site every 2 weeks, or monthly, the faithfulness of the limpets to their home scars can be assessed. Experience suggests that exposed or barnacle-dominated shore limpets are quite faithful, unlike those on smooth rock surfaces or under seaweeds in more sheltered conditions. Limpets can be individually marked by painted numbers or by using an epoxy resin to stick on plastic numbers (e.g. "Dynotape"). The shell has to be dry before applying the glue. First rinse off the shell with water and then dry the shell with alcohol.

All long-term studies need some way of marking or designating an area or a point. In many cases the topography of the shore can be used to provide landmarks. Sitings can be made from two or three landmarks using a compass to take bearings. The rock itself can be marked in various ways. Enamel paint lasts quite a long time (6 months to 2 years) if the rock is properly dried with a blow torch or by rubbing the rock with alcohol (methylated or rectified spirits are cheap). Rocks can be chiselled to form bench marks or markers. These can be made conspicuous and permanent markers by filling the chiselled groove or hole with ready mixed quick-set cement or epoxy resin. With epoxy resins the surface must be dry for bonding to occur.

Small drilled holes can be used as markers and can be made conspicuous by inserting coloured plastic plugs fixed with glue or by putting in a screw. Cordless drills are now quite cheap and excellent for field use. Drills driven by compressed air can be powered from SCUBA cylinders by adapting a two stage regulator (see Hawkins and Hartnoll, 1979); the complete set can be made for less than £150.

LONG TERM STUDIES AND EXPERIMENTS

Standard electric drills can be powered from a portable generator. A hand drill or brace and bit can be used in soft rocks. A clever way of finding markers in dense beds of seaweeds is to use a metal detector to find a fixed screw or bolt (Dr. G. Williams, pers comm.).

Field experiments

Rocky shores are ideal for field experiments. They are two dimensional, and are easy to sample non-destructively following manipulation. The underlying rationale is to set up treatments and untouched controls near to each other. In laboratory experiments every variable is kept constant except for the factor under investigation. In contrast, in field experiments all factors vary in parallel in the treatments and the controls except the variable manipulated. In recent years the use of field experiments has not only greatly increased our knowledge of the factors determining community structure on rocky shores, but has made a major contribution to general ecological theory (interested students should read Connell, 1972,1975 and the various chapters in Moore & Seed,1985).

If you have six months to two years available in which sites can be regularly visited then the scope for experimentation is considerable and will tell you more about the functioning of shore communities than descriptive studies – although descriptions are an essential preliminary to any experiment as first your hypothesis needs to be formulated.

Field experiments can explore the rôle of physical factors in setting distribution patterns by modifying shading, moisture or even rock topography such as crevice availability (e.g. Raffaelli and Hughes,1978) or by transplants (e.g. Schonbeck & Norton,1978,1980a; Norton,1985 for review). Alternatively, they can explore the rôle of biological interactions by various manipulations. For example, predators or grazers can be removed or excluded (e.g. Connell, 1961b; Southward, 1964; Paine, 1966; Lubchenco, 1978; Hawkins, 1981a; Hawkins & Hartnoll, 1983b). Competition can be investigated by manipulating densities and removing presumed competitors such as lower zoned barnacles (Connell, 1961a) or canopies of seaweeds (e.g. Schonbeck & Norton, 1980a; Hawkins & Harkin, 1985; Hawkins & Hartnoll, 1985; Norton, 1985 for review).

The following brief suggestions are two simple types of experiments which will yield rapid results and cause minimum damage to the shore communities. ***Careful consideration should be given to using techniques which avoid damage before starting any field experiment.*** It is essential to discuss any experiments with a lecturer or teacher before embarking on them. Consult your local marine laboratory to ensure that you are not using any sensitive areas or ruining monitoring or experimental areas.

Grazer exclusion or removal experiments

Hypothesis: Limpet grazing controls algal vegetation on exposed and moderately exposed shores (see Southward, 1964; Southward & Southward, 1978; Hawkins, 1981a; Hawkins & Hartnoll, 1983b).

Methods: Strictly speaking you should replicate these experiments. It is essential to do this if it is part of a research project – but if you are setting this up as a class demonstration, only one treatment and a control are necessary and less shore is damaged.

Choose an isolated rock or rocks covered by barnacles and limpets and surrounded by gulleys or crevices. An area of 1 x 1 m or 2 x 2 m is adequate. Select a similar area nearby at the same shore level as a control. Remove all limpets from the area you have designated as a treatment. Do not use too large an area because of the damage it will do to the shore community. The best time to start the experiment is January or February, but other times of the year are fine. The start time will affect the final results. Visit the area about every month and record the changes in algal cover in the control and treatment areas. A good series of photographs can be made. Remove any re-encroaching limpets in the treatment area. These can be measured and the rate of immigration assessed as supplementary information. On very exposed shores re-immigration will be more rapid and visits to remove the limpets may need to be made fortnightly.

Alternatively, anti-fouling paint makes an excellent barrier to limpet movement. Limpets do not like the taste and will not cross lines painted on the rocks. Any of the currently commercially available copper-based paints will do, particularly quick drying ones. Although at first glance painting the rock with a toxic agent does not seem a very conservation-minded thing to do, if applied sparingly and only surrounding a small area it will have little general effect. Again choose a barnacle covered surface. It is best to select a slightly convex surface to prevent toxic leachates running into the experimental areas (see Plate VIII, 15). Scrape off the barnacles in a 2-5 cm strip surrounding the experimental area. Scrub the strip using a wire brush. If a blow torch is available, lightly flame the surface to dry it.

An area of 0.5 x 0.5 m is often large enough; 1 x 1 m is fine and avoids the problem of leachates from the paint. Remove all the limpets from the treatment area. Set up two controls: one surrounded by a painted perimeter to control for the effects of paint and one without paint but marked in some way (e.g. 4 paint spots at the corners). Leave the limpets in place in each.

If a drill is available then little fences can be made to exclude the limpets. Small strips of chicken wire or galvanized square mesh can be fixed to the rock using screws anchored in plastic wall plugs. Epoxy resins can also be used to fix fences although drainage may be affected.

Again, visit the exclusions monthly (or fortnightly) and record changes in algal cover. The paint will need touching up and the odd encroaching limpet will need to be removed.

Results: Provided you have good removal and exclusion and there are few other grazers, such as topshells or winkles (which easily cross gullies or paint barriers), algae should proliferate in your removal or exclusion areas. The exact sequence will vary with season, shore level and even between years at the same place (see Hawkins and Hartnoll, 1983b for review). Usually, however, ephemeral species such as a brown film of diatoms, various green algae and *Porphyra* appear first, eventually followed by *Fucus*. Although in the summer it is possible for *Fucus* to colonize without any preliminary stages. Six months to a year will give good results so this can be done during the period of an 'A' level course or a college year.

Modifications: Comparisons of exclusions can be made between shore levels, in different seasons, with and without barnacle cover, wave exposure, rock type and angle of rock. Alternatively, different densities of limpets can be enclosed to find the numbers necessary to control algal growth on any particular shore.

LONG TERM STUDIES AND EXPERIMENTS

Fucus canopy removal experiments

Hypothesis: That shading and sweeping by the canopy of *Fucus* prevents growth of algae beneath it.

Methods: Choose an area of 1 x 1 m with a 90-100% *Fucus* canopy in either the *Fucus spiralis*, *F. vesiculosus* or *F. serratus* zone on a sheltered shore. **Do not use the Ascophyllum zone as recovery can take up to 20 years**. Clear the canopy by scraping away the holdfasts or by cutting the frond just above the holdfast. Garden shears are very good for this. If the plants are very long it is worth clearing a further 0.5-1 m buffer area surrounding the treatment. This will ensure that no sweeping of the treatment will occur. Choose an untouched control area nearby. Again if this is a research project it is essential to set up replicate treatments and controls. Two or three are probably sufficient and minimize damage to the community. An unreplicated experiment will be adequate only if a demonstration is required. Late winter or spring are the best time to remove the canopies. It is best to visit the experiment at monthly intervals to monitor changes.

Results: Depending on season and shore level, considerable colonization by ephemeral algae (mainly green) will occur within a month or so and the canopy will get back to 100% cover in between 1 and 3 years.

Modifications: Again, these experiments can be repeated at different seasons or at different levels within a zone. If the experiment is set up immediately beneath the zone of the species above then it is possible to demonstrate that competition sets the lower limit of zonation (see Schonbeck & Norton, 1980a). This is easiest with *Pelvetia* and *Fucus spiralis* as any *Fucus* sporelings can be selectively removed on subsequent visits (see Schonbeck & Norton, 1980a; Hawkins & Hartnoll, 1985). It is more difficult to distinguish between sporelings of different fucoids so repeated selective removal cannot easily be performed. In the spring the effects of canopy removal on barnacle settlement and survival can be investigated (Hawkins, 1983).

NOTES

CONSERVATION

In learning about the natural history and ecology of rocky shores it is important that we do not destroy them. Trampling by large numbers of students or pupils can disrupt communities. Think of the number of barnacles killed by each footstep. Destructive sampling on a large scale can devastate communities. The photographs in Plate IX, 1-3 show what can happen when limpets are removed. *Ascophyllum* canopies can take over 20 years to recover after harvesting or experimental removal. Therefore think very carefully before undertaking a study or unleashing a class of students. We have ourselves unwittingly damaged areas in the past by ill-thought out exercises and excursions. We now follow the guidelines below and hopefully have seen the error of our ways!

1)   **AVOID DESTRUCTIVE SAMPLING WHEREVER AND WHENEVER POSSIBLE**.

2)   **ALWAYS TRY AND IDENTIFY SPECIMENS AND ASSESS ABUNDANCE IN THE FIELD**. If you need to take organisms back to the laboratory remove one or two specimens only.

3)   **AVOID REMOVING LARGE NUMBERS OF SPECIMENS FOR MEASUREMENTS**. Make measurements in the field. If this is not possible, and only as a last resort, take them back to the laboratory and measure them and then return them to their approximate position on the shore on the next low tide. To keep animals alive keep them **cool**, and put them in moistened plastic trays or bags, with some damp *Fucus* on top. If you put them in seawater in bags they will rapidly deoxygenate or foul the seawater and die. Most intertidal animals can respire in air if kept moist. It is possible to do this with winkles, topshells and whelks. Most other animals are more sensitive. Limpets do not re-attach easily and usually die if removed.

4)   **DO NOT WRECK THE SHORE**. Always put back stones, or clumps of seaweed. The organisms sheltering underneath are usually delicate and may die if exposed to dry air and sunlight or be dislodged by waves or eaten by predators.

5)   **ALL TEACHERS SHOULD BE CAREFUL BEFORE LETTING LOOSE A LARGE CLASS ON THE SHORE**. Is the shore typical of many areas around and therefore of little conservation interest, or is it a rare rocky shore amongst depositing shores? Does the shore have a unique feature such as a *Sabellaria* reef? Discourage the inevitable tendency to kick off limpets or stamp on bladders of seaweeds. Think hard about whether your class is mature enough to appreciate the shore and respect the organisms present.

6)   **TRY TO AVOID HEAVY USAGE OF SPECIFIC AREAS**. It is a good idea to change locations from time to time, or use different parts of an extensive shore. If working at a Marine Research Laboratory consult with local scientists about sensitive areas or areas with work occurring on them. A field class can wreck a Ph.D. or a long term experiment.

7)   **BE VERY CAREFUL WHEN SETTING UP MANIPULATIVE EXPERIMENTS**. Always try and use as small an area and as few replicates as is possible without compromising your experiment. The less you disrupt the community, the more realistic your experiment will be.

Most of the above comments are common sense. Some shores and some organisms are more sensitive than others. Removing a small number of limpets from an exposed shore will cause little damage, there are plenty of juveniles to take the place of the adults and recruitment from the plankton is generally high. On a moderately sheltered shore this may not be the case and the ecological balance could be affected. Destructive sampling of species at their geographic limits (e.g. *Bifurcaria bifurcata, Cystoseira* spp., *Monodonta lineata, Paracentrotus, Balanus perforatus*) should be avoided: recruitment is likely to be low or highly variable and the population may not recover. Similarly, removal of animals like dogwhelks (*Nucella*) which do not have a planktonic stage will lead to the demise of populations as recruitment from elsewhere is very low and slow.

After so many don'ts, what can be done to encourage conservation of the marine environment? **The Marine Conservation Society** who commissioned this book take an active role in promoting interest in the sea and seashore, and also act as a pressure group on matters of marine nature conservation. If you are interested in joining them write to:

> The Marine Conservation Society
> 9 Gloucester Road
> Ross-on-Wye
> Herefordshire HR9 5BU

Ask for a membership form. They produce an excellent newsletter; meet once a year and provide a range of books on marine biology and marine conservation. Various recording schemes are in operation which they administer.

BIBLIOGRAPHY

This bibliography is not intended to be exhaustive but serves as a rapid way of finding available sources of reference. It will be particularly useful for teachers, undergraduates and possibly postgraduates. We have divided it into general ecological texts, identification guides and keys, and references to original papers and reviews. Many have been cited in the book. The others have been included for further reading and reference for the sake of completeness.

The Field Studies Council produces both papers and reviews on the ecology of common rocky shore plants and animals which are available for sale at reasonable prices. They also produce AIDGAP keys which undergo extensive testing before publication. These can be obtained from "Field Studies", Nettlecombe Court, Williton, Taunton, Somerset TA4 4HT, or the Richmond Publishing Company Ltd, Orchard Road, Richmond Surrey. They are indicated by the letters *FSC* at the end of the reference.

The Linnean Society produce the synopses of the British Fauna which are books on the general biology and identification of particular groups. These are more specialized and are aimed at advanced undergraduates, postgraduates and experienced scientists. The Marine Conservation Society produce identification guides based on colour prints (available from MCS Sales Ltd, 9, Gloucester Road, Ross-on-Wye, Herefordshire HR9 5BU). These are aimed at interested amateurs but are excellent for students. There is also a companion volume to this book primarily on sublittoral organisms (Guide to Inshore Marine Life, Erwin D. and Picton B. Immel).

General ecological texts

Barnes, R.S.K. and Hughes, R.N., 1982. An Introduction to Marine Ecology. Blackwell Scientific Publications. Excellent all round marine ecology book.

Carefoot, T.H., 1978. Pacific Seashores. J.J. Douglas Ltd. Vancouver. Although ostensibly on Pacific shores, it contains much information on Atlantic shores and general biology.

Lewis, J.R., 1964. The Ecology of Rocky Shores. English Universities Press Ltd. London. A comprehensive account of zonation patterns around the British Isles plus useful general information. The standard text on British shores.

Moore, P.G. & Seed, R. (eds), 1985. The Ecology of Rocky Coasts: essays presented to J.R. Lewis. Hodder and Stoughton. London. For the more advanced student, a compendium of reviews on various aspects of shore ecology to commemorate the retirement of J.R. Lewis.

Newell, R.C., 1979. Biology of Intertidal Animals (3rd. edition). Marine Ecological Surveys Ltd. A book looking primarily at the behaviour and physiology of intertidal organisms.

Southward, A.J., 1965. Life on the Seashore (The Scholarship Series in Biology). Heinemann Educational Books. London. Out of print but an interesting book outlining both sandy and rocky shores.

Stephenson, T.A. & Stephenson, A., 1972. Life Between Tidemarks on Rocky Shores. WH Freeman & Co., San Fransisco. This book allows British shores to be put into a worldwide perspective. A beautifully illustrated guide to zonation patterns worldwide.

Yonge, C.M., 1949. The Sea Shore. The New Naturalist, Collins, London (also Fontana paperback). Excellent account of the natural history of sea shores.

Checklists of Fauna and Flora

Bruce, J.R., Colman, J.S. & Jones, N.S., 1963. Marine Fauna of the Isle of Man. Liverpool University Press.

Howson, C.M., 1987. Directory of the British Marine Fauna and Flora. Marine Conservation Society, Ross-on-Wye.

Marine Biological Association, 1957. Plymouth Marine Fauna. (3rd edition). Plymouth.

South, G.R. & Titley, I., 1986. A checklist and distributional index of the benthic marine algae of the North Atlantic Ocean. Huntsman Marine Lab. and British Museum Natural History. St Andrews and London.

Methods and Statistics

Elliot, J.M., 1977. Some methods for the statistical analysis of samples of benthic invertebrates. Scientific Publications, Freshwater Biological Association, 25 (2nd edition).

Hiscock, K., 1985. Rocky Shore Survey and Monitoring Workshop. May 1st to 4th 1984. British Petroleum International Ltd, London.

Holme, N.A., and McIntyre, A.D., 1981. Methods for the study of Marine Benthos. (2nd edition), IPB Handbook 16, Blackwell Scientific Publications.

Meddis, R., 1975. Statistical handbook for non-statisticians. McGraw-Hill, London.

Price, J.H., Irvine, D.E.G. & Farnham, W.F., 1980. The Shore Environment, vol 1: Methods. Systematics Association, Academic Press, London.

Identification guides

General guides

Barrett, J.M. & Yonge, C.M., 1972. Pocket Guide to the Sea Shore (revised edition). Collins, London.

Campbell, A.C., 1976. Guide to the Seashore and Shallow Seas of Britain and Europe. Hamlyn, London.

Fish, J.D. & Fish, S.,1989. A Student's Guide to the Seashore. Unwin Hyman, London.

Hayward, P.J., 1988. Animals on seaweed. Naturalists handbook No.9. Richmond Publishing Co. Ltd.

Hayward, P.J., & Ryland, J.S., 1990. The Marine Fauna of the British Isles and North-west Europe. Vols I & II. Oxford University Press.

Quigley, M. & Crump, R., 1986. Animals and Plants of Rocky Shores. Blackwell Habitat Field Guides. Blackwell.

Specific groups classified by taxa

Lichens

Dobson, F.S., 1981. Lichens: An illustrated guide (2nd. revised edition). Richmond Publishing Co.

Duncan, U.K., 1959. A Guide to the Study of Lichens. Buncle, Arbroath.

Phillips, R., 1980. Grasses, Ferns, Mosses and Lichens of Great Britain and Ireland. Pan Books.

Seaweeds (Algae)

Clokie J.J.P. & Boney A.D.,1981. Green algae of the Firth of Clyde. Key to attached Chlorophyceae and some Xanthophyceae. Scottish Field Studies. 13-33.

Dickenson, C.I., 1963. British Seaweeds (The Kew Series). Eyre and Spottiswoode, London.

Hendey, N. I., 1964. An Introductory Account of the Smaller Algae of British Coastal Waters. Part V. Bacillariophyceae (Diatoms). MAFF Fisheries Investigations series IV. H.M.S.O.

Hiscock, S., 1979. A Field Key to the British Brown Seaweeds (Phaeophyta). Field Studies, 5:1-44. *FSC*

Hiscock, S., 1986. A Field Key to the British Red Seaweeds (Rhodophyta). Field Studies Council, Occasional Publications, 13. *FSC*

Jones, W.E., 1962. A Key to the Genera of British Seaweeds. Field Studies, 1(4):1-32.

Maggs, C. A. & Howson, C.M., 1985. A photographic guide to some common subtidal seaweeds of the British Isles. Marine Conservation Society, Ross-on-Wye. (Colour photographs plus text).

Newton, L., 1931. A handbook of the British Seaweeds. British Museum (Natural History), London.

Sponges (Porifera)

Ackers R.G., Moss D., Picton B.E. & Stone S.M.K., 1985. Sponges of The British Isles. Marine Conservation Society, Ross-on-Wye. (Colour photographs plus text).

Hydroids, Sea Anemones (Coelenterates or Cnidaria)

Hincks, T., 1868. A History of the British Hydroid Zoophytes. Vol. I and II. John Van Voorst, Paternoster Row. London.

Manuel, R.L., 1980. British Anthozoa. Synopses of the British Fauna (New Series), 18. The Linnean Society of London. Academic Press.

Manual, R.L., 1980. The Anthozoa of the British Isles. Marine Conservation Society, Ross-on-Wye. (Colour photographs plus text).

Flatworms (Platyhelminthes)

Prudhoe, S., 1982. British Polyclad Turbellarians. Synopses of the British Fauna (New Series), 26. The Linnean Society of London and The Estuarine and Brackish-Water Sciences Association. Cambridge University Press.

Ribbon worms (Nemertines)

Gibson, R., 1982. British Nemertines. Synopses of the British Fauna (New Series), 24. The Linnean Society of London and The Estuarine and Brackish-Water Sciences Association. Cambridge University Press.

Segmented Worms (Polychaetes)

Clark, R.B., 1960. Polychaete Fauna of the Clyde Sea Area. Scottish Marine Biological Association.

Garwood, P.R., 1981. The Marine Fauna of the Cullercoats District. No. 9. Polychaeta – Errantia. Dove Marine Laboratory, University of Newcastle-upon-Tyne.

Garwood, P.R., 1982. The Marine Fauna of the Cullercoats District. No. 10. Polychaeta-Sedentaria including Archiannelida. Dove Marine Laboratory, University of Newcastle-upon-Tyne.

George, J.D. & Hartmann-Schröder, G., 1983. Polychaetes: British Amphinomida, Spintherida and Eunicida. Synopses of the British Fauna (New Series), 32. The Linnean Society of London and The Estuarine and Brackish-Water Sciences Association. E.J. Brill.

Knight-Jones, P. & Knight-Jones, E.W., 1977. Taxonomy and ecology of British Spirobidae (Polychaeta). J. Mar. Biol. Ass. U.K., 57: 453-499.

BIBLIOGRAPHY

Sipunculids

Gibbs, P.E., 1977. British Sipunculans. Synopses of the British Fauna (New Series), 12. The Linnean Society of London. Academic Press.

Crustaceans

Crothers, J.H. & Crothers, M., 1983. A Key to the Crabs and Crab-like Animals of the British Inshore Waters. Field Studies, 5: 753-806. *FSC*

Holdich, D.M. & Jones, J.A., 1983. Tanaids. Synopses of the British Fauna (New Series), 27. The Linnean Society of London. Cambridge University Press.

Ingle, R.W., 1980. British Crabs. Oxford University Press, Oxford.

Ingle, R.W., 1983. Shallow-water Crabs. Synopses of the British Fauna (New Series), 25. The Linnean Society of London and The Estuarine and Brackish-Water Sciences Association. Cambridge University Press.

Lincoln, R.L., 1984. British Marine Amphipoda: Gammaridae. British Museum (Natural History) Publications, London.

Naylor, E., 1972. British Marine Isopods. Synopses of the British Fauna (New Series), 3. The Linnean Society of London. Academic Press.

Newman, W.A. & Ross, A., 1976. Revision of the balanomorph barnacles; including a catalog of species. Memoirs of the San Diego Society of Natural History, No. 9.

Southward, A.J., 1976. On the taxonomic status and distribution of *Chthamalus stellatus* (Cirripedia) in the north-west Atlantic region; with a key to the common intertidal barnacles of Britain. J. Mar. Biol. Ass. U.K., 56: 1007-1028.

Rainbow, P.S., 1984. An Introduction to the Biology of British Littoral Barnacles. Field Studies Council, 161. *FSC*

Smaldon, G., 1979. British Coastal Shrimps and Prawns. Synopses of the British Fauna (New Series), 15. The Linnean Society of London. Academic Press.

Sea spiders (Pycnogonids)

King, P.E., 1974. British Sea Spiders. Synopses of the British Fauna (New Series), 5. The Linnean Society of London. Academic Press.

Molluscs

Brown, G.H., and Picton, B.E., 1979. Nudibranchs of the British Isles – a colour guide. Underwater Conservation Society. (Colour photographs and text).

Graham, A., 1971. British Prosobranchs. Synopses of the British Fauna (New Series), 2. The Linnean Society of London. Academic Press.

Graham, A., 1988. Molluscs: Prosobranch and Pyramidellid Gastropods. Synopses of the British Fauna (New Series), 2. (2nd. edition). The Linnean Society of London and The Estuarine and Brackish-Water Sciences Association. E.J. Brill.

Jones, A.M. & Baxter, J.M., 1987. Molluscs; Caudofoveata, Solenogastres, Polyplacophora and Scaphopoda. Synopses of the British Fauna (New Series), 37. The Linnean Society of London and The Estuarine and Brackish-Water Sciences Association. E.J. Brill.

Tebble, N., 1976. British Bivalve Seashells (2nd edition). British Museum (Natural History) Publications, London.

Thompson, T.E., 1988. Molluscs: Benthic Opisthobranchs. Synopses of the British Fauna (New Series), 8 (2nd Ed.). The Linnean Society of London and The Estuarine and Brackish-Water Sciences Association. Academic Press.

Sea mosses or seamats (Bryozoans)

Hayward, P.J., 1985. Ctenostome Bryozoans. Synopses of the British Fauna (New Series), 33. The Linnean Society of London and The Estuarine and Brackish-Water Sciences Association. E.J. Brill.

Hayward, R.J. & Ryland, J.S., 1979. British Ascophoran Bryozoans. Synopses of the British Fauna (New Series), 14. The Linnean Society of London. Academic Press.

Hayward, R.J. & Ryland, J.S., 1985. Cyclostome Bryozoans. Synopses of the British Fauna (New Series), 34. The Linnean Society of London and The Estuarine and Brackish-Water Sciences Association. E.J. Brill.

Ryland, J.S. & Hayward, P.J., 1977. British Anascan Bryozoans. Synopses of the British Fauna (New Series), 10. The Linnean Society of London. Academic Press.

Sea urchins and starfish (Echinoderms)

Mortensen, T., 1927. Handbook of the British Echinoderms of the British Isles. Oxford University Press.

Picton, B.E., 1986. The Echinoderms of the British Isles. A colour Guide. Marine Conservation Society, Ross-on-Wye. (Colour photographs and text).

Seaquirts (Ascidians or tunicates)

Millar, R.H., 1970. British Ascidians. Synopses of the British Fauna (New Series), 1. The Linnean Society of London. Academic Press.

Picton, B.E., 1985. Ascidians of the British Isles. A Colour Guide. Marine Conservation Society, Ross-on-Wye. (Colour photographs and text).

Fishes

Dipper, F. A., 1987. British Sea Fishes. Underwater World Publications, London.

Lythgoe, J. & Lythgoe, G., 1971. Fishes of the Sea. Blandford, London.

Muus, B.J. & Dahlstrøm, P., 1974 (reprinted 1988). Sea Fishes of Britain and North-Western Europe. Collins.

Wheeler, A., 1978. Key to the Fishes of Northern Europe. Warne.

References (cited*)

Andrew, N. I. & Mapstone, B. D., 1987. Sampling and description of spatial pattern in marine ecology. Oceanogr. Mar. Biol. Ann. Rev., 25: 39-90.

Anon, 1974. Co-tidal and co-range lines – British Isles and adjacent waters. Hydrographer of the Navy, Taunton (Admiralty Chart 5058).

Anon, 1981. Atlas of the seas around the British Isles. Compiled by: Directorate of Fisheries Research, MAFF, Lowestoft.

*Baker, J. M. & Crothers, J.H., 1986. Intertidal Rock. Chapter 8. In: Biological Surveys of Estuaries and Coasts, eds J.M. Baker and W.J. Wolff. Estuarine and Brackish-Water Sciences Association handbook. Cambridge University Press.

*Ballantine, W.J., 1961. A biologically defined exposure scale for the comparative description of rocky shores. Field Studies I (3):1-17.

*Barnes, H. & Barnes, M., 1964. Some relations between the habitat, behaviour and metabolism on exposure to air of the high level intertidal cirrepede Chthamalus stellatus. Helgol. Wiss. Meeresunters, 10: 19-28.

Barnes, H. & Powell, H.T., 1950. The development, general morphology and subsequent elimination of barnacle populations, Balanus crenatus and B. balanoides, after a heavy initial settlement. J. Anim. Ecol., 19: 175-9.

*Baxter, J.M., Jones A.M., & Simpson, J. A., 1985. A study of long-term changes in some rocky shore communities in Orkney. Proc. Roy. Soc. Edinburgh, B75: 47-63.

Bennell, S. J., 1981. Some observations on the littoral barnacle populations of North Wales. Mar. Env. Res., 5: 227-240.

*Bertness, M.D., Yund, P.O. & Brown, A.F., 1983. Snail grazing and the abundance of algal crusts on a sheltered New England rocky beach. J. Exp. Mar. Biol. Ecol., 71 : 147-164.

*Boney, A.D., 1966. A Biology of Marine Algae. Hutchinson Educational Press.

Boney, A.D., 1979. Long term observations on the intertidal lichen Lichina pygmaea ag. J. Mar. Biol. Ass. U.K., 59: 801-803.

*Bowman, R.S. & Lewis, J.R., 1977. Annual fluctuations in the recruitment of Patella vulgata L. J. Mar. Biol. Ass. U.K., 57: 793-815.

*Bowman, R.S., 1981. The morphology of Patella spp. juveniles in Britain, and some phylogenetic influences. J. Mar. Biol. Ass. U.K., 61: 647-666.

*Branch, G.M., 1979. Aggression by limpets against invertebrate predators. Anim. Behaviour 27: 408 -410.

*Branch, G.M., 1981. The biology of limpets; physical factors, energy flow and ecological interactions. Oceanogr. Mar. Biol. Ann. Rev., 19: 235-379.

Branch, G. M. 1984. Competition between marine organisms: Ecological and evolutionary implications. Oceanogr. Mar. Biol. Ann. Rev., 22: 429-593.

*Bryan, G.W., Gibbs, P.E., Burt, G.R. & Hummerstone, L.G., 1987. The effects of Tributyltin (TBT) accumulation on adult dogwhelks, Nucella lapilus: long term field and laboratory experiments. J. Mar. Biol. Ass. U.K., 67: 525-544.

*Bryan, G.W., Gibbs, P.E., Hummerstone, L.G. & Burt, G.R., 1986. The decline of the gastropod Nucella lapillus around south west England: evidence for the effect of tributyltin from antifouling paints. J. Mar. Biol. Ass. U.K., 66: 611-640.

Burrows, E.M. & Lodge, S.M., 1950. Note on the interrelationships of Patella, Balanus and Fucus on a semi-exposed coast. Rep. Mar. Biol. Stn. Port Erin, 62: 30-34.

*Burrows, M.T., 1988. The comparative biology of Chthamalus stellatus Poli and Chthamalus montagui Southwood. Unpublished Ph.D. thesis, University of Manchester.

*Carter M.A. & Thorpe J.P.,1981. Reproductive, genetic and ecological evidence that Actinia equina var. mesembryanthemum and var. fragacia are not conspecific. J. Mar. Biol. Ass U.K., 61: 71-93.

Colman, J. S., 1933. The nature of intertidal zonation of plants and animals. J. Mar. Biol. Ass. U.K., 18: 435-476.

*Connell, J.H., 1961a. The influence of interspecific competition and other factors on the distribution of the barnacle Chthamalus stellatus. Ecol., 42: 710-723.

*Connell, J.H., 1961b. Effect of competition, predation by Thais lapillus and other factors on natural populations of the barnacle Balanus balanoides. Ecol. Monogr., 31: 61-104.

*Connell, J.H., 1972. Community interactions on marine intertidal rocky shores. Ann. Rev. Ecol. Syst., 3: 169-192.

*Connell, J.H.,1975. Some mechanisms producing structure in natural communities: a model and evidence from field experiments. Ch 16. In: Ecology and Evolution of Communities, ed. M.L. Cody & J.M. Diamond. Belknap Press, Cambridge, Mass., pp. 460-490.

Crapp, G.B., 1973. The distribution and abundance of animals and plants on the rocky shores of Bantry Bay. Irish Fish. Invest., B, 9: 1-35.

Crisp, D.J., 1964. The effects of the severe winter of 1962-1963 on marine life in Britain. J. Anim. Ecol., 33: 165-210.

Crisp, D. J., 1964. An assessment of plankton grazing by barnacles. In: Grazing in terrestrial and marine environments, ed D.J.Crisp. Blackwell Scientific Publications, Oxford. pp. 251-264.

*Crisp, D. J., 1974. Factors influencing the settlement of marine invertebrate larvae. In: Chemoreception in marine organisms, eds P.T.Grant & A.M.Mackie. Academic Press, London, New York. pp. 177-265.

*Crisp, D.J. & Southward, A.J., 1958. The distribution of intertidal organisms along the coasts of the English Channel. J. Mar. Biol. Ass. U.K., 37: 1037-1048.

BIBLIOGRAPHY

Crisp, D. J., Southward, A. J. & Southward, E. C., 1981. On the distribution of the intertidal barnacles *Chthamalus stellatus*, *Chthamalus montagui* and *Euraphia depressa*. J. Mar. Biol. Ass. U.K., 61: 359-380.

Cross, T.F. & Southgate, T., 1980. Mortalities of fauna of rocky substrates in south west Ireland associated with the occurrence of *Gyrodinium aureolum* blooms during autumn 1979. J. Mar. Biol. Ass. U.K., 60: 1071-1073.

*Crothers, J.H., 1975. On variation in *Nucella lapillus* L.: shell shape in populations from the south coast of England. Proc. Malac. Soc. Lond., 41: 489-498.

*Crothers, J.H. 1981. On the graphical presentation of quantitative data. Field Studies, 5: 487-511. *FSC*

*Crothers, J.H., 1985. Dogwhelks: An introduction to the biology of *Nucella lapillus*. Field Studies, 6: 291-360. *FSC*

*Crothers, J.H., 1987. On the graphical presentation of quantitative data: a postscript. Field Studies, 6: 685-693. *FSC*

*Crothers, J.H., 1989. Has the population decline due to TBT pollution affected shell-shape variation in the dog-whelk, *Nucella lapillus* (L.)? J. Moll. Stud., 55: 461-467.

*Dalby, D.H., Cowell, E.B., Syratt, W.J. & Crothers, J.H., 1978. An exposure scale for marine shores in western Norway. J. Mar. Biol. Ass. U.K., 58: 975-96.

Dayton, P. K.,1971. Composition disturbance and community organisation. The provision and subsequent utilisation of space. Ecol. Monogr., 41: 351-389.

Dayton, P. K., 1975. Experimental evaluation of ecological dominance in a rocky intertidal algal community. Ecol. Monogr., 45: 137-159.

*Denny, M.W., 1983. A simple device for measuring the maximum force exerted on intertidal organisms. Limnol. Oceanogr., 28: 1269-1274.

Dethier, M. N. 1984. Disturbance and recovery in intertidal pools: Maintenance of mosaic patterns. Ecol. Monogr., 54: 99-118.

*Doty, M.S., 1971. Measurement of water movement in reference to benthic algal growth. Botanica Mar., 14: 32-35.

Dring, M.J., 1982. The Biology of Marine Plants. Edward Arnold, London.

*Ebling, F.J., Sloane, J.F., Kitching, J.A. & Davies, H.M.,1962. The ecology of lough Ine. XII. The distribution and characteristics of *Patella* species. J. Anim. Ecol. 31, 457-470.

Ebling, F. J., Kitching, J. A., Muntz, L. & Taylor, C. M., 1964. The ecology of Lough Ine. XIII. Experimental observations of the destruction of *Mytilus edulis* and *Nucella lapillus* by crabs. J. Anim.Ecol., 33: 73-82.

*Foster, B.A., 1971. On the determinants of the upper limit of intertidal distribution of barnacles (Crustacea: Cirrepeda). J. Anim. Ecol., 40: 33-48.

*Fretter, V. & Graham, A, 1962. British Prosobranch Molluscs. Ray Society, London.

Gaines, S. D. & Lubchenco, J., 1982. A unified approach to marine plant-herbivore interactions. II. Biogeography. Ann. Rev. Ecol. Syst., 13: 111-138.

Gaines, S. D. & Roughgarden, J.,1985. Larval settlement rate: a leading determinant of structure in an ecological community of the intertidal zone. Proc. Natl. Acad. Sci. U.S.A., 82: 3707-3711.

*Gibbs, P.E., Bryan, G.W., Pascoe, P.L. & Burt, G.R.,1987. The use of the dogwhelk *Nucella lapillus* as an indicator of T.B.T. contamination. J. Mar. Biol. Ass. U.K., 67: 507-523.

*Gosling, E.M., 1984. The systematic status of *Mytilus galloprovincialis* in western Europe: a review. Malacologia, 25: 551-568.

*Gruet, Y., 1986. Spatio-temporal changes of Sabellarian reefs built by the sedentary polychaete *Sabellaria alveolata* (Linné). Mar. Ecol. (Naples), 7: 289-372.

Grubb, V. M.,1936. Marine algal ecology and the exposure factor at Peveril Point, Dorset. J.Ecol., 24: 392-423.

*Hartnoll, R. G., 1983. Bioenergetics of a limpet grazed intertidal community. S. Afr. J. Sci., 79: 116-117.

Hartnoll, R.G. & Hawkins, S.J., 1980. Monitoring rocky shore communities: a critical look at spatial and temporal variation. Helgol. Wiss. Meeresunters, 33: 484-495.

*Hartnoll, R.G. & Hawkins, S.J., 1985a. Patchiness and fluctuations on moderately exposed rocky shores. Ophelia, 24 (1): 53-63.

Hartnoll, R.G. & Hawkins, S.J., 1985b. The use of multiple random quadrats for the assessment of the abundance of rocky shore organisms. In: Rocky Shore Survey and Monitoring Workshop, ed K. Hiscock, BP International, London, pp. 73-74.

*Hartnoll, R.G. and Wright J. R., 1977. Foraging movements and homing in the limpet *Patella vulgata* L. Animal Behaviour, 25: 806-810.

*Hawkins, S.J., 1981a. The influence of *Patella* grazing on the fucoid/barnacle mosaic on moderately exposed rocky shores. Kieler Meeresforsch., 5: 537-543.

*Hawkins, S.J., 1981b. The influence of season and barnacles on the algal colonization of *Patella vulgata* exclusion areas. J. Mar. Biol. Ass. U.K., 61: 1-15.

*Hawkins, S.J., 1983. Interactions of *Patella* and macroalgae with settling *Semibalanus balanoides* (L.). J. Exp. Mar. Biol. Ecol., 71: 55-72.

*Hawkins, S.J., 1985. Simple statistical methods for use in surveys and monitoring of rocky shores. In: Rocky Shore Survey and Monitoring Workshop, ed. K. Hiscock. BP International, London. pp. 105-117.

*Hawkins, S.J. & Hartnoll, R.G.,1979. A compressed air drill powered by SCUBA cylinders for use on rocky shores. Estuar. Cstl. Mar. Sci., 9: 819-820.

Hawkins, S.J. & Hartnoll, R.G., 1980. Small scale relationships between species number and area on rocky shores. Estuar. Cstl. Mar. Sci., 10: 201-214.

*Hawkins, S.J. & Hartnoll, R.G., 1982a. Settlement patterns of *Balanus balanoides* (L.) in the Isle of Man. J. Exp. Mar. Biol. Ecol., 62: 271-283.

*Hawkins, S.J. & Hartnoll, R.G., 1982b. The influence of barnacle cover on the numbers growth and behaviour of *Patella vulgata* on a vertical pier. J. Mar. Biol. Ass. U.K., 62: 865-867.

*Hawkins, S.J. & Hartnoll, R.G., 1983a. Changes in a rocky shore community: an evaluation of monitoring. J. Mar. Env. Res., 9: 131-181.

*Hawkins, S.J. & Hartnoll, R.G., 1983b. Grazing of intertidal algae by marine invertebrates. Oceanogr. Mar. Biol. Ann. Rev., 21: 195-282.

*Hawkins, S.J. & Hartnoll, R.G., 1985. Factors controlling the upper limits of intertidal canopy algae. Mar. Ecol. Prog. Ser., 20: 265-271.

*Hawkins, S.J. & Harkin, E., 1985. Experimental canopy removal in algal dominated communities low on the shore and in the shallow sub-tidal. Botanica Mar., 28: 223-230.

Hawkins, S. J., Watson, D., Hill, A. S., Harding, S. P., Kyriakides, M. A., Hutchinson, S. & Norton, T. A., 1989. A comparison of feeding mechanisms in microphagous herbivorous intertidal prosobranchs in relation to resource partitioning. J. Moll. Stud., 55: 151-165.

*Haylor, G.S., Thorpe, J.P. & Carter, M.A., 1984. Genetic and ecological differentiation between sympatric colour morphs of the common intertidal sea anemone *Actinia equina*. Mar. Ecol., 16: 281-289.

Hill, A. S. & Hawkins, S.J., 1990. Methods for the investigation of microbial films on rocky shores. J. Mar. Biol. Ass. U.K., 70: 77-88.

*Hughes, R.N. & Dunkin, S. de B.,1984a. Behavioural components of prey selection by dogwhelks, *Nucella lapillus* (L.) feeding on mussels *Mytilus edulis* (L.) in the laboratory. J. Exp. Mar. Biol. Ecol., 77: 45-68.

*Hughes, R.N. & Dunkin, S. de B., 1984b. Effect of dietary history on selection of prey, and foraging behaviour among patches of prey, by the dogwhelk, *Nucella lapillus* (L.). J. Exp. Mar. Ecol., 79: 159-172.

*Hughes, R.N. & Elner, R.W., 1979. Tactics of a predator, *Carcinus maenas*, and morphological responses of the prey, *Nucella lapillus*. J. Anim. Ecol., 48: 65-78.

*Imrie, D.W., Hawkins, S.J. & McCrohan, C.R., 1989. The olfactory-gustatory basis of food preference in the herbivorous prosobranch, *Littorina littorea* L. J. Moll.Stud., 55: 217-225.

*Jones, H.D., 1985. Shell cleaning behaviour in *Callistoma zizyphinum*. J. Moll. Stud., 50, 245-247.

*Jones, H.G. & Norton, T.A., 1979. Internal factors controlling the rate of evaporation from fronds of some intertidal algae. New Phytol., 83: 771-781.

*Jones, H.G. & Norton, T.A., 1981. The role of internal factors in controlling evaporation from intertidal algae. In: Plants and their atmospheric environment, ed. J. Grace, E.D. Ford & P.G. Jarvis. Blackwell Scientific Publishers, Oxford, Brit. Ecol. Soc. Symp., No. 21, pp. 231-235.

*Jones, N. S., 1948. Observations and experiments on the biology of *Patella vulgata* at Port St. Mary, Isle of Man. Proc.Trans. Lpool Biol. Soc., 56: 60-77.

*Jones, W.E. & Demetropoulos, A., 1968. Exposure to wave action: measurements of an important ecological parameter on rocky shores on Anglesey. J. Exp. Mar. Biol. Ecol., 2: 46-63.

*Kain, J.M., 1979. A view of the genus *Laminaria*. Oceanogr. Mar. Biol. Ann. Rev., 17: 101-161.

*Kendall, M.A., 1988. The age and size structure of some Northern populations of the trochid gastropod *Monodonta lineata*. J. Moll. Studies, 53 (2): 213-222.

*Kendall, M.A., Bowman, R.S., Williamson, P. & Lewis, J.R., 1985. Annual variation in the recruitment of *Semibalanus balanoides* on the north Yorkshire coast, 1969-1981. J. Mar. Biol. Ass. U.K., 65: 1009-1030.

Kitching, J. A. & Ebling, F. J., 1961. The ecology of Lough Ine. XI. The control of algae by *Paracentrotus lividus* (Echiniodea). J. Anim. Ecol., 30: 373-383.

*Kitching, J.A., Muntz, L. & Ebling, F., 1966. The ecology of Lough Ine, XV. The ecological significance of shell and body forms of *Nucella*. J. Anim. Ecol., 35: 113-126.

Knight, S.J.T. & Mitchell, R., 1980. The survey and nature conservation of littoral areas. In: The Shore Environment, vol 1: Methods, eds J.H. Price, D.E.G. Irvine & W.F. Farnham. Systematics Association, Academic Press, London, pp. 303-21.

*Knight-Jones E.W., 1951. Gregariousness and some other aspects of the settling behaviour of spirorbids. J. Mar. Biol. Ass. U.K. 30: 201-222.

*Knight-Jones E.W., 1953. Laboratory experiments on gregariousness during settling in *Balanus balanoides* and other barnacles. J. Exp. Biol., 30: 584-598.

Lewin, R., 1986. Supply-side ecology. Science, 234: 25-27.

Lewis, J.R., 1976. Long-term ecological surveillance: practical realities in the rocky littoral. Oceanogr. Mar. Biol. Ann. Rev., 14: 371-390.

*Lewis, J.R., 1977. The role of physical and biological factors in the distribution and stability of rocky shore communities. In: Biology of Benthic Organisms, eds B.F. Keegan, P.O. Ceidigh & P.J.S. Boaden, Pergamon Press, Oxford. pp. 417-424.

Lewis, J. R. & Bowman, R. S., 1975. Local habitat-induced variations in the population dynamics of *Patella vulgata* L. J. Exp. Mar. Biol. Ecol., 17: 165-203.

*Little,C., 1989. Factors governing patterns of foraging activity in littoral marine herbivorous molluscs. J. Moll. Stud., 55: 273-284.

Little, C. & Smith, L.P., 1980. Vertical zonation on rocky shores in the Severn Estuary. Estuar. Cstl. Mar. Sci., 11: 651-669.

*Little C., Williams, G.A., Morrit, D., Perrins, J.M., & Stirling P.E.,1988. Foraging behaviour of the limpet *Patella vulgata* L. in an Irish Sea Lough. J. Exp. Mar. Biol. Ecol., 120:1-21.

BIBLIOGRAPHY

Le Tourneux, F. & Bourget, E., 1988. Importance of physical and biological settlement cues used at different spatial scales by the larvae of *Semibalanus balanoides*. Mar. Biol., 97: 57-66.

*Lodge, S.M., 1948. Algal growth in the absence of *Patella* on an experimental strip of foreshore, Port St Mary, Isle of Man. Proc. Trans. Lpool Biol. Soc., 56: 78-83.

Lowell, R.B., 1984. Desiccation of intertidal limpets: effects of shell size, fit to substratum, and shape. J. Exp. Mar. Biol. Ecol., 77: 197-207.

*Lubchenco, J., 1978. Plant species diversity in a marine intertidal community: importance of herbivore food preference and algal competitive abilities. Am. Nat., 112: 23-39.

Lubchenco, J., 1980. Algal zonation in the New England rocky intertidal community: an experimental analysis. Ecology, 61: 333-344.

Lubchenco, J., 1983. *Littorina* and *Fucus*: effects of herbivores, substratum heterogeneity, and plant escapes during succession. Ecology, 64: 1116-1123.

Lubchenco, J. & Cubit, J. D., 1980. Heteromorphic life histories of certain marine algae as adaptation to variations in herbivory. Ecology, 61: 676-686.

Lubchenco, J. & Gaines, S. D., 1981. A unified approach to marine plant-herbivore interaction. I. Populations and Communities. Ann. Rev. Ecol. Syst., 12: 405-437.

Lubchenco, J. & Menge, B. A., 1978. Community development and persistence in a low rocky intertidal zone. Ecol. Monogr., 48: 67-94.

Lumb, C.M., 1985. Survey for the assessment of nature conservation importance of sites. In: Rocky Shore Survey and Monitoring Workshop, ed. K. Hiscock, BP International, London, pp. 24-32.

*Menge, B. A., 1976. Organisation of the New England rocky intertidal communities: roles of predation, competition, environmental heterogeneity. Ecol. Monogr., 46: 355-393.

Menge, B. A.,1978. Predation intensity in a rocky intertidal community. Oecologia (Berlin), 34: 1-16.

Menge, B. A. & Lubchenco, J., 1981. Community organisation in temperate and tropical rocky intertidal habitats. Prey refuges in relation to consumer pressure gradients. Ecol. Monogr., 51: 429-450.

Menge, B. A. & Sutherland, J. P., 1976. Species diversity gradients: synthesis of the roles of predation, competition and spatial heterogeneity. Am. Nat., 110: 351-369.

Menge, B. A. & Sutherland, J. P., 1987. Community regulation: variation in disturbance, competition and predation in relation to environmental stress and recruitment. Am. Nat., 130: 730-757.

*Moore, H.B., 1934. The relation of shell growth to environment in *Patella vulgata*. Proc. Malac. Soc. Lond., 21: 217-222.

*Muus, B.J., 1968. A field method for measuring 'exposure' by means of plaster balls. A preliminary account. Sarsia, 34: 61-68.

Naylor, E., 1985. Tidally rhythmic behaviour of marine animals. In: Symposia of the Society for Experimental Biology, XXXIX. Physiological adaptations of marine animals, ed. M.S. Laverack. The Company of Biologists, Cambridge, pp.63-93.

Nelson-Smith, A., 1965. Marine Biology of Milford Haven: the physical environment. Field. Studies., 2: 155-188.

*Newell, R.C., 1979. Biology of Intertidal Animals (3rd. edition). Marine Ecological Surveys Ltd.

*Norton, T.A.,1985. The zonation of seaweeds on rocky shores. In: Moore, P.G. and Seed, R. The Ecology of Rocky Coasts. pp. 7-21. Hodder and Stoughton.

Norton, T.A., 1985b. Provisional Atlas of the Marine Algae of Britain & Ireland, ed. for the British Phycological Society by T.A. Norton, Institute of terrestrial Ecology, Huntington.

Norton, T. A., Hawkins, S. J., Manley, N. J., Williams, G. A. & Watson, D. C., 1990. Scraping a living: A review of littorinid grazing. Hydrobiologica, In press.

*Paine, R.T.,1966. Food web complexity and species diversity. Am. Nat. 100: 65-75.

*Paine, R.T.,1974. Intertidal community structure: experimental studies on the relationship between a dominant competitor and its principal predator. Oecologia (Berlin), 15: 93-120.

Paine, R.T., 1980. Food webs: linkage, interaction, strength and community infrastructure. J. Anim. Ecol., 49: 667-685.

*Palumbi, S.R., 1984. Measuring intertidal wave forces. J. Exp. Mar. Biol. Ecol., 81: 171-179.

Petraitis, P.S., 1983. Grazing patterns of the periwinkle *L. littorea* and their effect on intertidal organisms. Ecology, 64: 522-523.

Pyefinch, K. A., 1943. The intertidal ecology of Bardsey Island, North Wales with special reference to the recolonisation of rocky shores, and the rock pool environment. J. Anim. Ecol., 12: 82-108.

*Quicke, D.L.J. & Brace, R.C., 1983. Phenotypic and genotypic spacing within an aggregation of the sea anemone, *Actinia equina*. J. Mar. Biol. Ass. U.K., 63: 493-515.

*Quicke, D.L.J. & Brace, R.C., 1984. Evidence for the existence of a third, ecologically distinct morph of the anemone, *Actinia equina*. J. Mar. Biol. Ass. U.K., 64: 531-534.

*Quicke, D.L.J., Donoghue, A.M. & Brace, R.C., 1983. Biochemical genetic and ecological evidence that red/brown individuals of the anemone *Actinia equina* comprise two morphs in Britain. Mar. Biol., 77: 29-37.

*Raffaelli, D.G., 1982. Recent ecological research on some European species of *Littorina*. J. Moll. Stud., 48: 342-354.

Raffaelli, D.G., 1985. Functional feeding groups of some intertidal molluscs defined by gut contents analysis. J. Moll. Stud., 51: 233-239.

*Raffaelli, D.G. and Hughes, R.N., 1978. The effect of crevice size and availability on populations of *Littorina rudis* and *Littorina neritoides*. J. Anim. Ecol. 47: 71-83.

*Rugg, D.A. and Norton, T.A., 1987. *Pelvetia canaliculata*, a high-shore seaweed that shuns the sea. In, Plant Life in Aquatic and Amphibious Habitats, ed. R.M.M. Roberts. Brit. Ecol. Soc. Symp., No. 5 pp. 347-358.

*Russell, G., 1973. The 'litus' line: a re-assessment. Oikos, 24: 158 - 61.

*Russell, G. & Fielding, A.H., 1981. Individuals, populations and communities. In: The Biology of Seaweeds, eds C.S. Lobban & M.J. Wynne. Blackwell Scientific Publications, Oxford, pp.393-420.

*Schonbeck, M.W. & Norton, T.A., 1978. Factors controlling the upper limits of fucoid algae on the shore. J. Exp. Mar. Biol. Ecol., 31: 303-313.

*Schonbeck, M.W. & Norton, T.A., 1979a. An investigation of drought avoidance in intertidal fucoid algae. Botanica Mar., 22: 133-144.

Schonbeck, M.W. & Norton, T.A., 1979b. The effects of brief periodic submergence on intertidal fucoid algae. Estuar. Coastal. Mar. Sci., 8: 205-211.

*Schonbeck, M.W. & Norton, T.A., 1979c. Drought hardening in the upper-shore seaweeds *Fucus spiralis* and *Pelvetia caraliculata*. J. Ecol., 67: 687-696.

*Schonbeck, M.W. & Norton, T.A., 1980a. Factors controlling the lower limits of fucoid algae on the shore. J. Exp. Mar. Biol. Ecol., 43: 131-150.

*Schonbeck, M.W. & Norton, T.A., 1980b. The effects on intertidal fuciod algae of exposure to air under various conditions. Botanica Mar., 23: 141-147.

*Seed, R., 1969. The ecology of *Mytilus edulis* (L.) (Lamellibranchiata) on exposed rocky shores. II. Growth and mortality. Oecologia (Berlin), 3: 317-350.

Shanks, A. L. & Wright, W. G., 1986. Adding teeth to wave action: the destructive effects of wave born rocks on intertidal organisms. Oecologia (Berlin), 69: 420-428.

*Sole-Cava, A.M. & Thorpe, J.P., 1987. Further genetic evidence for the reproductive isolation of green sea anemone *Actinia prasina* Gosse from common intertidal beadlet anenome *Actinia equina* (L.) Mar. Ecol., 38: 225-229.

Sousa, W. P., 1979a. Experimental investigations of disturbance and ecological succession in a rocky intertidal algal community. Ecol. Monogr., 49: 227-254.

Sousa, W. P., 1979b. Disturbance in marine intertidal boulder fields: the non-equilibrium maintenance of species diversity. Ecology, 60: 1225-1239.

Sousa, W. P., 1980. The responses of a community to a disturbance: the importance of successional age and species life histories. Oecologia (Berlin), 45: 72-81.

Sousa, W. P., 1984a. The role of disturbance in natural communities. Ann. Rev. Ecol. Syst., 15: 353-391.

Sousa, W. P., 1984b. Intertidal mosaics: patch size, propagule availability and spatially variable patterns of succession. Ecology, 65: 1918-1935.

Sousa, W. P., 1985. Disturbance and patch dynamics on rocky intertidal shores. In: The Ecology of Natural Disturbance and Patch Dynamics, eds S. T. A. Pickett & P. S. White. Academic press, pp. 101-124.

Southward, A.J., 1953. The ecology of some rocky shores in the south of the Isle of Man. Proc. Trans. Lpool. Biol.Soc., 59: 1-50.

Southward, A. J., 1956. The population balance between limpets and seaweeds on wave-beaten rocky shores. Rep. Mar. Biol. Stn. Port Erin, 68: 20-29.

Southward, A.J., 1958. The zonation of animals and plants on rocky sea shores. Biol. Rev., 33: 137-177.

Southward, A.J., 1964. Limpet grazing and the control of vegetation on rocky shores. In: Grazing in Terrestrial and Marine Environments, ed D.J. Crisp. Blackwell, Oxford, pp.265-273.

Southward, A.J., 1976. On the taxonomic status and distribution of *Chthamalus stellatus* (Cirripedia) in the north-east Atlantic region with a key to the common intertidal barnacles of Britain. J. Mar. Biol. Ass. U.K., 56: 1007-1028.

Southward, A. J. & Crisp, D. J., 1954. Recent changes in the distribution of the intertidal barnacles *Chthamalus stellatus* (Poli) and *Balanus balanoides* (L.) in the British Isles. J. Anim.Ecol., 23:163-177.

*Southward, A.J. & Southward, E.C., 1978. Recolonisation of rocky shores in Cornwall after use of toxic dispersants to clean up the Torrey Canyon spill. J. Fish. Res. Bd. Can., 35: 682-706.

*Thomas, M.L.H., 1986. A physically derived exposure index for marine shorelines. Ophelia, 25: 1-13.

Thompson, G.B., 1979. Distribution and population dynamics of the limpet *Patella aspera* Lamark in Bantry Bay. J. Exp. Mar. Biol. Ecol., 40: 115-135.

Thompson, G.B., 1980. Population dynamics of the limpet *Patella vulgata* (L.) in Bantry Bay. J. Exp. Mar. Biol. Ecol., 45: 173-217.

*Todd, C.D. & Lewis, J.R., 1984. Effects of low air temperature on *Laminaria digitata* in south west Scotland. Mar. Ecol. Prog. Ser.,16, pp. 199-201.

*Underwood, A.J., 1979. The ecology of marine gastropods. Adv. Mar. Biol., 16: 111-210.

*Underwood, A.J.,1980. The effects of grazing by gastropods and physical factors on the upper limits of of distribution of intertidal macroalgae. Oecologia (Berlin), 46: 201-213.

Underwood, A. J., Denley, E. J. & Moran, M. J.,1983. Experimental analyses of the structure and dynamics of mid-shore rocky intertidal communities in New South Wales. Oecologia (Berlin), 56: 202-219.

*Underwood, A. J. & Jernakoff, P., 1981. Interactions between algae and grazing gastropods in the structure of a low-shore algal community. Oecologia (Berlin), 48: 221-233.

Underwood, A. J. & Jernakoff, P., 1984. The effects of tidal height, wave exposure, seasonality and rock pools on intertidal macroalgae. J. Exp. Mar. Biol. Ecol., 75: 71-96.

Underwood, A. J. & Fairweather, P. G., 1989. Supply-side ecology and benthic marine assemblages. Trends in Ecology and Evolution, 4: 16-20.

Vadas, R. L., 1985. Herbivory. In: Handbook of Phycological Methods. Ecological Field Methods: Macroalgae, eds M. M. Littler & D. S.Littler. Cambridge University Press, Cambridge, pp. 531-572.

*Watson, D.C. & Norton, T.A., 1985a. Dietary preferences of the common periwinkle *Littorina littorea*. J. Exp. Mar. Biol. Ecol., 88: 193-211.

*Watson, D.C. & Norton, T.A., 1985b. The physical characteristics of seaweed thalli as deterrents to littorine grazers. Botanica Mar., 28: 383-387.

*Watson, D.C. and Norton, T.A., 1987. The habitat and feeding preferences of *Littorina obtusata* (L.) and *Littorina mariae* Sacchi et Rastelli. J. Exp. Mar. Biol. Ecol., 112: 61-72.

*Williams, G.A., 1987. Niche partitioning in *Littorina obtusata* (L.) and *L. mariae* Sacchi et Rastelli. Unpublished Ph.D. thesis, University of Bristol.

*Williamson, P. & Kendall, M.A., 1981. Population age structure and growth of the trochid *Monodonta lineata* determined from shell rings. J. Mar. Biol. Ass. U.K., 61: 1011-1026.

*Wilson, D.P., 1971. *Sabellaria* colonies at Duckpool, North Cornwall, 1961-70. J. Mar. Biol. Ass. U.K., 51: 509-580.

Wolcott, T.G., 1973. Physiological ecology and intertidal zonation in limpets (*Acmaea*): a critical look at 'limiting factors'. Biol. Bull. Mar. Biol. Lab., Woods Hole, 145: 389-422.

*Wright, R.A.D.,1981. Wave Exposure Studies on Rocky Shores in Shetland. Unpublished Ph.D. thesis, University of London (Imperial College of Science and Technology).

Wright, J.R. & Hartnoll, R.G., 1981. An energy budget for a population of the limpet *Patella vulgata*. J. Mar. Biol. Ass. U.K., 61: 627-646.

NOTES

GLOSSARY

Abundance scales. Scales describing the abundance of organisms. Groupings are in several broad categories covering a the range from rare to abundant.

Aerobic. With air.

Algal film. Growth of small microscopic algae covering rocks in the euphotic zone. It makes the rock slippery!

Anoxic/Anaerobic. Without oxygen/without air.

Biological exposure scale. Scale devised by Ballantine (1961) to define wave exposure based on the abundance of key species.

Biological interaction. When one organism affects another through competition, grazing, predation or parasitism.

Biomass. The weight of all organisms forming a specific population or community in a given area.

Biota. Living organisms (animals or plants).

Brown algae or Phaeophyta. Algae with brown secondary pigments.

Browsers. Animals that scrape the thin layer of living organisms off the surface of the substratum.

Boreal. Northern or arctic species.

Boundary layer. Thin layer next to the surface of an object with different flow characteristics.

Calcareous. Organisms which have calcium carbonate (chalk) as part of their structure.

Canopy cover. Proportion of a specified area consisting of canopy species which form a layer above other biota.

Canopy species. Algae which grow taller than other algal species and therefore receive the majority of light by forming a layer or canopy.

Chart datum. Set reference point on admiralty charts for water depth in relation to tides.

Community. The plants and animals living together in a given area.

Competition. When one individual (**intra**specific competition) or species (**inter**specific competition) interacts with another individual or species and has a deleterious effect. Competition is usually seen to be for some resource in limited supply such as food, or in the intertidal, space.

Cyanobacteria. Previously called cyanophytes or blue green algae, now considered to be blue green bacteria because of simple cellular structure lacking an organized nucleus.

Density. Number of organisms occurring within a specified area.

Destructive sampling. Organisms are removed from the natural environ-ment for assessment in the laboratory, and therefore part of the shore community is destroyed.

Detritus. Dead or dying organisms, often in fragments which are con-sumed by detritivores and broken down by bacteria.

Dichotomous keys. Identification aid based on a series of either-or-choices between morphological features.

Diploid. Chromosomes occur in homologous pairs so that twice the haploid number are present to give the full genetic complement. The usual pattern in most adult organisms.

Double tides. Tides where a double peak occurs at high water or low water usually due to complex patterns of water movement (e.g. Portland to Portsmouth on south coast). The double low water can result in a long stand of water when the tide goes out.

Dry weight. Weight of an organism after moisture has been removed from the tissues.

Ebb-tide. Outgoing or falling tide.

Encrusting plants. Non-erect plants forming a sheet over the substrate surface.

Energetics and energy flow. The process whereby energy is transformed (within individuals or in the case of flow, communities) following primary production.

Environmental gradient. A region where the environment is changing in a particular direction.

Ephemeral algae. Short-lived, fast-growing, opportunistic seaweeds.

Eulittoral. Midshore region of the shore between the littoral fringe and sublittoral fringe defined by the presence of barnacles in exposed conditions, fucoids in shelter.

Exposure. The degree of wave action on a shore governed by the distance of open sea over which the wind may blow to generate waves (**the fetch**) and the strength and incidence of the winds. **Exposed usually refers to wave action not aerial exposure.**

Filter feeder. An animal which obtains food by filtering it from the water column. Sometimes called **suspension** feeders.

Flood-tide. Rising tide.

Foraging. Feeding excursion.

Fucoids. Wracks, large, tough, leathery brown seaweeds which are typical of the intertidal area of rocky shores.

Gametes. Haploid reproductive cells (sperms and eggs) whose nuclei and sometimes cytoplasm fuse in fertilization to form a diploid zygote.

Gametophyte. Seaweed in haploid gamete-producing phase.

Green algae or Chlorophyta. Algae with only green pigments.

Haploid. Organism or reproductive cell (gamete) with a single set of unpaired chromosomes (c.f. diploid)

Hermaphrodite. Animal which can be both sexes either at the same time or sequentially. E.g. limpets change from being males to females with age.

Littoral fringe. Area of very high shore with lower limit demarked by the top of the barnacle zone and the upper limit of winkle species and black lichens.

Microhabitat. A small part of the shore which has different physical conditions e.g. the open rock, a rock pool, crevice, on or under seaweed.

Mobile. Organisms with the ability to move, often very slowly, on rocky shores.

Neap tide. Smallest tide, least difference between high and low water.

Non-destructive sampling. Sampling without removing organisms from the environment and altering the community.

Osmosis. The directional passage of water across a semi-permeable membrane in response to a concentration gradient. It results in water moving from dilute to concentrated conditions.

Ovate. Shaped like an egg.

Percentage cover. Proportion of rock surface or canopy space occupied by organisms.

Perennial turf-forming algae. Long lived algae which form turfs over the substrate.

Plankton. Organisms which drift in the water and have limited powers of locomotion in comparison with the horizontal water movements. Many shore animals have larvae in the plankton which act as a dispersive phase.

Polymorphic. A species which has several different genetically controlled forms, usually colouring, but can be shell shape or sculpture.

Population. The plants or animals of a particular species in a set area.

Primary production. The production of new living material by plant photosynthesis. Productivity is the rate of production in a given area.

Quadrat. Boundary of a rectangular area being sampled. Now used to mean the frame used to delimit this area.

Red algae or Rhodophyta. Algae with red secondary pigments.

Resource. A commodity used by an organism such as light in plants, food in animals, or habitat space.

Salinity. The concentration of dissolved salts in seawater.

Sessile. Organisms that are fixed to the substratum.

Sheltered. Low wave action area such as a bay or harbour.

Sporophyte. The phase in an algal life history producing diploid spores, e.g. the large *Laminaria* plants.

Spring tide. Largest tide, greatest difference between high and low water.

Station. A position on a transect where a sample or set of samples are taken.

Stress gradient. An environmental gradient along which physiological stress can be defined; e.g. with increasing tidal height, aerial exposure will increase.

Symbiotic. An association between two organisms, usually considered to involve mutal benefits.

Sublittoral fringe. Lowest zone of the intertidal which is usually only uncovered during spring tides, dominated by kelps or red algae.

Substrate cover. Amount of rock covered by sessile or encrusting forms.

Succession. The sequence of colonization over time following the creation of new space in a community. **NOT the change in community composition along an environmental gradient.** Succession tends to proceed to an end point or climax community, except where continual disturbance restarts the sequence.

Taxonomy. The study and process of naming and classifying organisms.

Thallus. The frond of an alga. As algae do not have roots all exchange with the environment occurs through the thallus including nutrient uptake.

Transect. A section down or across a shore (or any other habitat) along which observations or experiments are made.

Understorey. Organisms occurring under the main canopy of algae.

Weathering. The process whereby rock is eroded.

Wet weight. Gross weight of an organism, without removing any water from the tissues.

Zoospores. Motile spores possessing flagella, present in certain algae and fungi which are produced within a sporangium.

Zygotes. Fertilized eggs before they undergo cleavage.